Ornamental Fish Aquaculture

Ornamental Fish Aquaculture

– Author –
Ahtesham Malik

Wordpen Academics
Publishing Science That Matter

ISBN: 978-93-90438-11-2 (Hardback)

Published by : **WORDPEN ACADEMICS**
4736/23, Ansari Road, Darya Ganj,
New Delhi – 110 002
Phone: +91-011-35004336
E-mail: info@wordpenacademics.com

Printed at **:** **Replika Press Pvt. Ltd.**

Exclusive Distributor: ***Bio-Green Books, New Delhi***

Preface

Aquarium fishes are called ornamental fishes. Ornamental fish rearing is becoming a common hobby among many people all over the world. It has taken a permanent place in the public places like parks, aquaria, hotels, offices, etc. A small aquarium placed at a suitable place in the human habitat gives so much joy and refreshment to the minds of all who see it. At present most people buy ornamental fishes from the professional breeders and sellers and rear them. But in many cases the ornamental fishes breed in the aquarium indiscriminately causing degeneration of the breed and at times inviting diseases connected with inbreeding. Breeding ornamental fish has become a commercial activity by itself. For carrying out proper knowledge is necessary. This booklet intends to provide some basic ideas about breeding of ornamental fishes to anyone who is interested in it. Ornamental fish keeping is one of the most popular hobbies in the world today. The growing interest in aquarium fishes has resulted in steady increase in aquarium fish trade globally. The trade with a turnover of US $ 5 Billion and an annual growth rate of 8 percent offers a lot of scope for development.

The top exporting country is Singapore followed by Hong Kong, Malaysia, Thailand, Philippines, Sri Lanka, Taiwan, Indonesia and India. The largest importer of Ornamental fish is the USA followed by Europe and Japan. The emerging markets are China and South Africa. Over US $ 500 million worth of ornamental fish are imported into the USA each year. Ninety five per cent of our ornamental fish export is based on wild collection. Majority of the indigenous ornamental fish trade in India is from the North Eastern states and the rest is from Southern states which are the hot spots of fish bio diversity in India. This capture based export is not sustainable and it is a matter of concern for the industry.

In order to sustain the growth it is absolutely necessary to shift the focus from capture to culture based development. Moreover most of the fish species grown for their ornamental importance can be bred in India successfully. Organised trade in ornamental fish depends on assured and adequate supply of demand, which is possible only by mass breeding. There are quite a large number of tropical aquarium fishes known to aquarists. While many of the fish are easy to breed, some of these are rare, difficult to breed and expensive. Most of the exotic species can be bred and reared easily since the technology is simple and well developed. The major expense of any animal rearing is the expenses on feeding. Fishes at various stages also require different types of feed as the protein and energy requirement for different categories of fishes is varying. Hence we can have fry feed, fingerling feed, grower feed, breeder stock feed, spawner feed, etc. The younger the fish greater the need for the protein requirement. Accordingly the amount of nutrients in the feed as well as the amount of ingredients in the feed should be changed. Such feeds can be bought or prepared oneself at least to some extent. Breeding of the ornamental fishes is a science as well as an art which one has to master through gaining knowledge and skill from the experience of others as well as one's own. The general behaviour and breeding pattern of each species is different form another. Hence for successful breeding one has to learn in detail the species characteristics in its behaviour and breeding habits. Keeping ornamental fishes is not only a hobby but also a commercial activity for many people in our country. For many people it has become a profession. Indeed to manage professional aquarium professionals are needed to do the job satisfactorily. This booklet, though only dealing with the subject at an elementary level, will help the beginners to acquire sufficient knowledge about breeding of ornamental fishes.

This comprehensive book will inspire and catalyse fishery scientists to undertake in depth research and extension activities and would prove to be an essential compendium for researchers, research institutions and planners.

—*Ahtesham Malik*

Contents

1

Introduction

Ornamental Fish

Carp, along with many of their cyprinid relatives, are popular ornamental aquarium and pond fish. The two most notable ornamental carps are goldfish and koi. Goldfish and koi have advantages over most other ornamental fishes, in that they are tolerant of cold (they can survive in water temperatures as low as 4°C), can survive at low oxygen levels, and can tolerate low water quality.

Figure: An unusual goldfish breed: An oranda-type variegated pearlscale.

Goldfish (*Carassius auratus*) were originally domesticated from the Prussian carp (*Carassius gibelio*), a dark greyish-brown carp native to Asia. They were first bred for colour in China over a thousand years ago. Due to selective breeding, goldfish have been developed into many distinct breeds, and are found in various colours, colour patterns, forms and sizes far different from those of the original carp. Goldfish were kept as ornamental fish in China for hundreds of years before being introduced to Japan in the 15th century, and to Europe in the late 17th century. Koi are a domesticated variety of common carp (*Cyprinus carpio*) that have been selectively bred for colour. The common carp was introduced from China to Japan, where selective breeding of the common carp in the 1820s in the Niigata region resulted in koi. In Japanese culture, koi are treated with affection, and seen as good luck. They are popular in other parts of the world as outdoor pond fish.

As Sport Fish

Carp Fishing: Izaak Walton said about carp in *The Compleat Angler*, "The Carp is the queen of rivers; a stately, a good, and a very subtle fish; that was not at first bred, nor hath been long in England, but is now naturalised."

Carp are variable in terms of angling value:

- In Europe, even when not fished for food, they are eagerly sought by anglers, being considered highly prized coarse fish that are difficult to hook. The UK has a thriving carp angling market. It is the fastest growing angling market in the UK, and has spawned a number of specialised carp angling publications such as 'Carpology', 'advanced carp fishing', 'carpworld', 'TotalCarp' and karper, and many informative carp angling web sites, such as Carpfishing UK and Carpit Carp Anglers Bulgaria
- In the United States, carp are also classified as a rough fish, as well as damaging to naturalised exotic species, but with sporting qualities. Many states' departments of natural resources are beginning to view the carp as an angling fish instead of a maligned pest. Groups such as the Carp Anglers Group and American Carp Society promote the sport and work with fisheries departments to organise events to introduce and expose others to the unique opportunity the carp offers freshwater anglers.

Types of Carp

1. Genus *Abramis*
   - Common bream or carp bream (*Abramis brama*)
2. Genus *Barbodes*:
   - Carnatic carp (*Barbodes carnaticus*)
3. Genus *Carassius*:
   - Crucian carp (*Carassius carassius*)
   - Prussian carp or Gibel carp (*Carassius gibelio*)
   - Goldfish (*Carassius auratus*)
4. Genus *Cirrhinus*:
   - Chinese mud carp (*Cirrhinus chinensis*)
   - Deccan white carp (*Cirrhinus fulungee*)

- Hora white carp (*Cirrhinus macrops*)
- Small-scale mud carp (*Cirrhinus microlepis*)
- Mud carp (*Cirrhinus molitorella*)
- Man carp (*Cirrhinus saprian*)

5. Genus *Ctenopharyngodon*:
   - Grass carp (*Ctenopharyngodon idellus*)
6. Genus *Culter*:
   - Predatory carp (*Culter erythropterus*)
7. Genus *Cyprinus*:
   - Common carp, European carp, koi (*Cyprinus carpio*)
8. Genus *Epalzeorhynchos*:
   - Red-tailed black shark (*Epalzeorhynchos bicolour*)
   - Rainbow shark or red-finned black shark (*Epalzeorhynchos frenatus*)
9. Genus *Henicorhynchus*:
   - Siamese mud carp (*Henicorhynchus siamensis*)
10. Genus *Hypophthalmichthys*:
    - Bighead carp (*Hypophthalmichthys nobilis*)
    - Silver carp (*Hypophthalmichthys molitrix*)
11. Genus *Labeo*:
    - African carp (*Labeo coubie*)
    - Fringed-lipped peninsula carp (*Labeo fimbriatus*)
    - Pigmouth carp (*Labeo kontius*)
12. Genus *Mylopharyngodon*:
    - Black carp (*Mylopharyngodon piceus*)

Disease in Ornamental Fish

Ornamental fish kept in aquariums are susceptible to numerous diseases. The study of fish diseases has remained a rudimentary branch of veterinary medicine. Due to their generally small size and the low cost of replacing diseased or dead fish, the cost of testing and treating diseases is often seen as more trouble than the value of the fish.

Issues in Diagnosis and Treatment

Due to the artificially limited volume of water and high concentration of fish in most aquarium tanks, communicable diseases often affect most or all fish in a tank. An improper nitrogen cycle, inappropriate aquarium plants and potentially harmful freshwater invertebrates can directly harm or add to the stresses on ornamental fish in a tank. Despite this, many diseases in captive fish can be avoided or prevented through proper water conditions and a well-adjusted ecosystem within the tank.

Prevention

Disease cures are almost always more expensive and less effective than simple prevention measures. Often precautions involve maintaining a stable aquarium that is adjusted for the specific species of fish that are kept and not over-crowding a tank or over-feeding the fish. Common preventive strategies include avoiding the introduction of infected fish, invertebrates or plants by quarantining new additions before adding them to an established tank, and discarding water from external sources rather than mixing it with clean water. Similarly, foods for herbivorous fish such as lettuce or cucumbers should be washed before being placed in the tank. Containers that do not have water filters or pumps to circulate water can also increase stress to fish. Other stresses on fish and tanks can include certain chemicals, soaps and detergents, and impacts to tank walls causing shock waves that can damage fish.

Treatment

In some cases the causes of an infection or disease will be obvious (such as fin rot), though in other cases it may be due to water conditions, requiring special testing equipment and chemicals to appropriately adjust the water. Isolating diseased fish can help prevent the spread of infection to healthy fish in the tank. This also allows the use of chemicals or drugs which may damage the nitrogen cycle, plants or chemical filtration of a properly-functioning tank. Other alternatives include short baths in a bucket that contains the treated water. Salt baths can be used as an antiseptic and fungicide, and will not damaging beneficial bacteria, though ordinary table salt may contain additives which can harm fish. Alternatives include aquarium salt, Kosher salt or rock salt. Gradually raising the temperature of the tank may kill certain parasites, though some diseased fish may be harmed and certain species can not tolerate high temperatures. Aeration is necessary since less oxygen is dissolved in warm water.

Fish Diseases and Parasites

Like humans and other animals, fish suffer from diseases and parasites. Fish defences against disease are specific and non-specific. Non-specific defences include skin and scales, as well as the mucus layer secreted by the epidermis that traps microorganisms and inhibits their growth. If pathogens breach these defences, fish can develop inflammatory responses that increase the flow of blood to infected areas and deliver white blood cells that attempt to destroy the pathogens. Specific defences are specialised responses to particular pathogens recognised by the fish's body, that is adaptative immune responses. In recent years, vaccines have become widely used in aquaculture and ornamental fish, for example vaccines for furunculosis in farmed salmon and koi herpes virus in koi. Some commercially important fish diseases are VHS, ich and whirling disease.

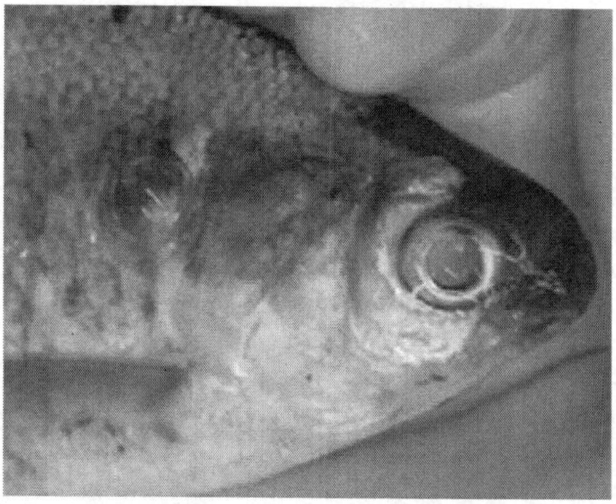

Figure: This gizzard shad has VHS, a deadly infectious disease which causes bleeding. It afflicts over 50 species of freshwater and marine fish in the northern hemisphere.

Disease

All fish carry pathogens and parasites. Usually this is at some cost to the fish. If the cost is sufficiently high, then the impacts can be characterised as a disease. However disease in fish is not understood well. What is known about fish disease often relates to aquaria fish, and more recently, to farmed fish.

Disease is a prime agent affecting fish mortality, especially when fish are young. Fish can limit the impacts of pathogens and parasites with behavioural or biochemical means, and such fish have reproductive advantages. Interacting factors result in low grade infection becoming fatal diseases. In particular, things that causes stress, such as natural droughts or pollution or predators, can precipitate outbreak of disease.

Disease can also be particularly problematic when pathogens and parasites carried by introduced species affect native species. An introduced species may find invading easier if potential predators and competitors have been decimated by disease.

Pathogens which can cause fish diseases comprise:

- viral infections
- bacterial infections, such as *Pseudomonas fluorescens* leading to fin rot and fish dropsy

- fungal infections
- water mould infections, such as *Saprolegnia* sp.
- metazoan parasites, such as copepods
- unicellular parasites, such as *Ichthyophthirius multifiliis.*

Parasites

Parasites in fish are a natural occurrence and common. Parasites can provide information about host population ecology. In fisheries biology, for example, parasite communities can be used to distinguish distinct populations of the same fish species co-inhabiting a region. Additionally, parasites possess a variety of specialised traits and life-history strategies that enable them to colonise hosts. Understanding these aspects of parasite ecology, of interest in their own right, can illuminate parasite-avoidance strategies employed by hosts.

Usually parasites (and pathogens) need to avoid killing their hosts, since extinct hosts can mean extinct parasites. Evolutionary constraints may operate so parasites avoid killing their hosts, or the natural variability in host defensive strategies may suffice to keep host populations viable. Parasite infections can impair the courtship dance of male three spine sticklebacks. When that happens, the females reject them, suggesting a strong mechanism for the selection of parasite resistance."

However not all parasites want to keep their hosts alive, and there are parasites with multistage life cycles who go to some trouble to kill their host. For example, some tapeworms make some fish behave in such a way that a predatory bird can catch it. The predatory bird is the next host for the parasite in the next stage of its life cycle. Specifically, the tapeworm *Schistocephalus solidus* turns infected threespine stickleback white, and then makes them more buoyant so that they splash along at the surface of the water, becoming easy to see and easy to catch for a passing bird.

Other parasitic disorders, include *Gyrodactylus salaris*, *Ichthyophthirius multifiliis*, cryptocaryon, velvet disease, *Brooklynella hostilis*, Hole in the head, *Glugea*, *Ceratomyxa shasta*, *Kudoa thyrsites*, *Tetracapsuloides bryosalmonae*, *Cymothoa exigua*, leeches, nematode, flukes, *Platyhelminthes*, carp lice and salmon lice

Cleaner Fish

Some fish take advantage of cleaner fish for the removal of external parasites. The best known of these are the Bluestreak cleaner wrasses of the genus *Labroides* found on coral reefs in the Indian Ocean and Pacific Ocean. These small fish maintain so-called "cleaning stations" where other fish, known as hosts, will congregate and perform specific movements to attract the attention of the cleaner fish. Cleaning behaviours have been observed in a number of other fish groups, including an interesting case between two cichlids of the same genus, *Etroplus maculatus*, the cleaner fish, and the much larger *Etroplus suratensis*, the host.

More than 40 species of parasites may reside on the skin and internally of the ocean sunfish, motivating the fish to seek relief in a number of ways. In temperate regions, drifting kelp fields harbour cleaner wrasses and other fish which remove parasites from the skin of visiting sunfish. In the tropics, the *mola* will solicit cleaner help from reef fishes. By basking on its side at the surface, the sunfish also allows seabirds to feed on parasites from their skin. Sunfish have been reported to breach more than ten feet above the surface, possibly as another effort to dislodge parasites on the body.

Mass Die Offs

Some diseases result in mass die offs. One of the more bizarre and recently discovered diseases produces huge fish kills in shallow marine waters. It is caused by the ambush predator dinoflagellate *Pfiesteria piscicida*. When large numbers of fish, like shoaling forage fish, are in confined situations such as shallow bays, the excretions from the fish encourage this dinoflagellate, which is not normally toxic, to produce free-swimming zoospores. If the fish remain in the area, continuing to provide nourishment, then the zoospores start secreting a neurotoxin. This toxin results in the fish developing bleeding lesions, and their skin flakes off in the water. The dinoflagellates then eat the blood and flakes of tissue while the affected fish die. Fish kills by this dinoflagellate are common, and they may also have been responsible for kills in the past which were thought to have had other causes. Kills like these can be viewed as natural mechanisms for regulating the population of exceptionally abundant fish. The rate at which the kills occur increases as organically polluted land runoff increases.

Wild Salmon

According to Canadian biologist Dorothy Kieser, protozoan parasite *Henneguya salminicola* is commonly found in the flesh of salmonids. It has been recorded in the field samples of salmon returning to the Queen Charlotte Islands. The fish responds by walling off the parasitic infection into a number of cysts that contain milky fluid. This fluid is an accumulation of a large number of parasites.

Henneguya and other parasites in the myxosporean group have a complex lifecycle where the salmon is one of two hosts. The fish releases the spores after spawning. In the Henneguya case, the spores enter a second host, most likely an invertebrate, in the spawning stream. When juvenile salmon out-migrate to the Pacific Ocean, the second host releases a stage infective to salmon. The parasite is then carried in the salmon until the next spawning cycle. The myxosporean parasite that causes whirling disease in trout, has a similar lifecycle. However, as opposed to whirling disease, the *Henneguya* infestation does not appear to cause disease in the host salmon — even heavily infected fish tend to return to spawn successfully.

According to Dr. Kieser, a lot of work on *Henneguya salminicola* was done by scientists at the Pacific Biological Station in Nanaimo in the mid-1980s, in particular, an overview report which states that "the fish that have the longest fresh water residence time as juveniles have the most noticeable infections. Hence in order of prevalence coho are most infected followed by sockeye, chinook, chum and pink." As well, the report says that, at the time the studies were conducted, stocks from the middle and upper reaches of large river systems in British Columbia such as Fraser, Skeena, Nass and from mainland coastal streams in the southern half of B.C. "are more likely to have a low prevalence of infection." The report also states "It should be stressed that *Henneguya*, economically deleterious though it is, is harmless from the view of public health. It is strictly a fish parasite that cannot live in or affect warm blooded animals, including man".

According to Klaus Schallie, Molluscan Shellfish Program Specialist with the Canadian Food Inspection Agency, "*Henneguya salminicola* is found in southern B.C. also and in all species of salmon. I have previously examined smoked chum salmon sides that were riddled with cysts and some sockeye runs in Barkley Sound (southern B.C., west coast of Vancouver Island) are noted for their high incidence of infestation."

Sea lice, particularly *Lepeophtheirus salmonis* and a variety of *Caligus* species, including *Caligus clemensi* and *Caligus rogercresseyi*, can cause deadly infestations of both farm-grown and wild salmon. Sea lice are ectoparasites which feed on mucous, blood, and skin, and migrate and latch onto the skin of wild salmon during free-swimming, planktonic *naupli* and *copepodid* larval stages, which can persist for several days. Large numbers of highly populated, open-net salmon farms can create exceptionally large concentrations of sea lice; when exposed in river estuaries containing large numbers of open-net farms, many young wild salmon are infected, and do not survive as a result. Adult salmon may survive otherwise critical numbers of sea lice, but small, thin-skinned juvenile salmon migrating to sea are highly vulnerable. On the Pacific coast of Canada, the louse-induced mortality of pink salmon in some regions is commonly over 80%.

Farmed Salmon

In 1972, Gyrodactylus, a monogenean parasite, spread from Norwegian hatcheries to wild salmon, and devastated some wild salmon populations.

In 1984, infectious salmon anemia (ISAv) was discovered in Norway in an Atlantic salmon hatchery. Eighty percent of the fish in the outbreak died. ISAv, a viral disease, is now a major threat to the viability of Atlantic salmon farming. It is now the first of the diseases classified on List One of the European Commission's fish health regime. Amongst other measures, this requires the total eradication of the entire fish stock should an outbreak of the disease be confirmed on any farm. ISAv seriously affects salmon farms in Chile, Norway, Scotland and Canada, causing major economic losses to infected farms. As the name implies, it causes severe anemia of infected fish. Unlike mammals, the red blood cells of fish have DNA, and can become infected with viruses. The fish develop pale gills, and may swim close to the water surface, gulping for air. However, the disease can also develop without the fish showing any external signs of illness, the fish maintain a normal appetite, and then they suddenly die. The disease can progress slowly throughout an infected farm and, in the worst cases, death rates may approach 100 percent. It is also a threat to the dwindling stocks of wild salmon. Management strategies include developing a vaccine and improving genetic resistance to the disease.

Figure: Atlantic salmon

In the wild, diseases and parasites are normally at low levels, and kept in check by natural predation on weakened individuals. In crowded net pens they can become epidemics. Diseases and parasites also transfer from farmed to wild salmon populations. A recent study in British Columbia links the spread of parasitic sea lice from river salmon farms to wild pink salmon in the same river." The European Commission (2002) concluded "The reduction of wild salmonid abundance is also linked to other factors but there is more and more scientific evidence establishing a direct link between the number of lice-infested wild fish and the presence of cages in the same estuary." It is reported that wild salmon on the west coast of Canada are being driven to extinction by sea lice from nearby salmon farms. Antibiotics and pesticides are often used to control the diseases and parasites.

Aquarium Fish

Ornamental fish kept in aquariums are susceptible to numerous diseases. In most aquarium tanks, the fish are at high concentrations and the volume of water is limited. This means that communicable diseases can spread rapidly to most or all fish in a tank. An improper nitrogen cycle, inappropriate aquarium plants and potentially harmful freshwater invertebrates can directly harm or add to the stresses on ornamental fish in a tank. Despite this, many diseases in captive fish can be avoided or prevented through proper water conditions and a well-adjusted ecosystem within the tank. Ammonia poisoning is a common disease in new aquariums, especially when immediately stocked to full capacity. Due to their generally small size and the low cost of replacing diseased or dead aquarium fish, the cost of testing and treating diseases is often seen as more trouble than the value of the fish.

Spreading Disease and Parasites

The capture, transportation and culture of bait fish can spread damaging organisms between ecosystems, endangering them. In 2007, several American states, including Michigan, enacted regulations designed to slow the spread of fish diseases, including viral hemorrhagic septicemia, by bait fish. Because of the risk of transmitting *Myxobolus cerebralis* (whirling disease), trout and salmon should not be used as bait. Anglers may increase the possibility of contamination by emptying bait buckets into fishing venues and collecting or using bait improperly. The transportation of fish from one location to another can break the law and cause the introduction of fish and parasites alien to the ecosystem.

Eating Raw Fish

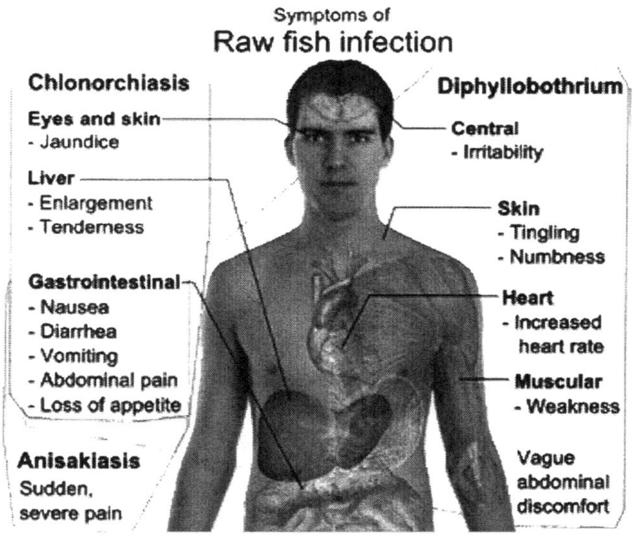

Figure: Differential symptoms of parasite infection by raw fish: Clonorchis sinensis (a trematode/fluke), Anisakis (a nematode/roundworm) and Diphyllobothrium a (cestode/ tapeworm), all have gastrointestinal, but otherwise distinct, symptoms.

Though not a health concern in thoroughly cooked fish, parasites are a concern when human consumers eat raw or lightly preserved fish such as sashimi, sushi, ceviche, and gravlax. The popularity of such raw fish dishes makes it important for consumers to be aware of this risk. Raw fish should be frozen to an internal temperature of "20°C ("4°F) for at least 7 days to kill parasites. It is important to be aware that home freezers may not be cold enough to kill parasites. Traditionally, fish that live all or part of their lives in fresh water were considered unsuitable for sashimi due to the possibility of parasites. Parasitic infections from freshwater fish are a serious problem in some parts of the world, particularly Southeast Asia. Fish that spend part of their life cycle in salt water, like salmon, can also be a problem. A study in Seattle, Washington showed that 100% of wild salmon had roundworm larvae capable of infecting people. In the same study farm raised salmon did not have any roundworm larvae.

Parasite infection by raw fish is rare in the developed world (fewer than 40 cases per year in the U.S.), and involves mainly three kinds of parasites: Clonorchis sinensis (a trematode/fluke), Anisakis (a nematode/roundworm) and Diphyllobothrium (a cestode/tapeworm). Infection risk of anisakis is particularly higher in fishes which may live in a river such as salmon (*shake*) in Salmonidae, mackerel (*saba*). Such parasite infections can generally be avoided by boiling, burning, preserving in salt or vinegar, or freezing overnight. Even Japanese people never eat raw salmon and ikura, and even if they seem raw, these foods are not raw but are frozen overnight to prevent infections from parasites, particularly anisakis.

Carp

Carp are various species of oily freshwater fish of the family Cyprinidae, a very large group of fish native to Europe and Asia. The cypriniformes (family Cyprinidae) are traditionally grouped with the Characiformes, Siluriformes and Gymnotiformes to create the super order Ostariophysi, since these groups have certain common features, such as being found predominantly in fresh water and that they possess Weberian ossicles (an anatomical structure originally made up of small pieces of bone formed from four or five of the first vertebrae); the most anterior bony pair is in contact with the extension of the labyrinth and the posterior with the swimbladder. The function is poorly understood, but this structure is presumed to take part in the transmission of vibrations from the swimbladder to the labyrinth and in the perception of sound, which would explain why the Ostariophysi have such a great capacity for hearing.

Most cypriniformes have scales and teeth on the inferior pharyngeal bones which may be modified in relation to the diet. *Tribolodon* is the only cyprinid genus which tolerates salt water, although there are several species which move into brackish water, but return to fresh water to spawn. All of the other cypriniformes live in continental waters and have a wide geographical range.

Some consider all cyprinid fishes carp, and the family Cyprinidae itself is often known as the carp family. In colloquial use, however, carp usually refers only to several larger cyprinid species such as Cyprinus carpio (common carp), Carassius carassius (Crucian carp), Ctenopharyngodon idella (grass carp), Hypophthalmichthys molitrix (silver carp), and Hypophthalmichthys nobilis (bighead carp). Carp have long been an important food fish to humans, as well as popular ornamental fishes such as the various goldfish breeds and the domesticated common carp variety known as koi. As a result, carp have been introduced to various locations, though with mixed results. Several species of carp are listed as invasive species by the U.S. Department of Agriculture, and worldwide large sums of money are spent on carp control.

Aquaculture

Various species of carp have been domesticated and reared as food fish across Europe and Asia for thousands of years. These various species appear to have been domesticated independently, as the various domesticated carp species are native to different parts of Eurasia. For example, the common carp, *Cyprinus carpio*, is originally from Central Europe. Several carp species (collectively known as Asian carp) were domesticated in East Asia.

Carp that are originally from South Asia, for example catla (*Gibelion catla*), rohu (*Labeo rohita*) and mrigal (*Cirrhinus cirrhosus*), are known as Indian carp. Their hardiness and adaptability have allowed domesticated species to be propagated all around the world.

Demand has declined, partly due to the appearance of more desirable table fish such as trout and salmon through intensive farming, and environmental constraints.

However, fish production in ponds is still a major form of aquaculture in Central and Eastern Europe, including the Russian Federation, where most of the production comes from low or intermediate-intensity ponds. In Asia, the farming of carp continues to surpass the total amount of farmed fish volume of intensively sea-farmed species, such as salmon and tuna.

Fish Farming

Fish farming is the principal form of aquaculture, while other methods may fall under mariculture. Fish farming involves raising fish commercially in tanks or enclosures, usually for food. A facility that releases young (juvenile) fish into the wild for recreational fishing or to supplement a species' natural numbers is generally referred to as a fish hatchery. The most common fish species raised by fish farms are salmon, carp, tilapia, European seabass, catfish and cod.

There is an increasing demand for fish and fish protein, which has resulted in widespread overfishing in wild fisheries. Fish farming offers fish marketers another source. However, farming carnivorous fish, such as salmon, does not always reduce pressure on wild fisheries, since carnivorous farmed fish are usually fedfishmeal and fish oil extracted from wild forage fish. In this way, the salmon can consume in weight more wild fish than they weigh themselves. The global returns for fish farming recorded by the FAO in 2008 totalled 33.8 million tonnes worth about $US 60 billion.

Major Categories of Fish Aquaculture

There are two kinds of aquaculture: extensive aquaculture based on local photosynthetical production and intensive aquaculture, in which the fish are fed with external food supply.

Extensive Aquaculture

Limiting for growth here is the available food supply by natural sources, commonly zooplankton feeding on pelagic algae or benthic animals, such as crustaceans and mollusks.Tilapia species filter feed directly on phytoplankton, which makes higher production possible.

The photosynthetic production can be increased by fertilizing the pond water with artificial fertilizer mixtures, such as potash, phosphorus, nitrogen and micro-elements.

Because most fish are carnivorous, they occupy a higher place in the trophic chain and therefore only a tiny fraction of primary photosynthetic production (typically 1%) will be converted into harvest-able fish.

A second point of concern is the risk of algal blooms. When temperatures, nutrient supply and available sunlight are optimal for algal growth, algae multiply their biomass at an exponential rate, eventually leading to an exhaustion of available nutrients and a subsequent die-off.

The decaying algal biomass will deplete the oxygen in the pond water because it blocks out the sun and pollutes it with organic and inorganic solutes (such as ammonium ions), which can (and frequently do) lead to massive loss of fish.

Another option is to use a wetland system such as that of Veta La Palma. In order to tap all available food sources in the pond, the aquaculturist will choose fish species which occupy different places in the pond ecosystem, e.g., a filter algae feeder such as tilapia, a benthic feeder such as carp or catfish and a zooplankton feeder (various carps) or submerged weeds feeder such as grass carp.

Despite these limitations significant fish farming industries use these methods. In the Czech Republic thousands of natural and semi-natural ponds are harvested each year for trout and carp. The large ponds around Trebon were built from around 1650 and are still in use.

Intensive Aquaculture

Optimal water parameters for cold- and warm-water fish in intensive aquaculture Acidity pH 6-9

Arsenic	<440 µg/L
Alkalinity	>20 mg/L (as $CaCO_3$)
Aluminum	<0.075 mg/L
Ammonia (non-ionised)	<O.O2mg/L
Cadmium	<0.0005 mg/L in soft water;<0.005 mg/L in hard water
Calcium	>5 mg/L
Carbon dioxide	<5-10 mg/L
Chloride	>4.0 mg/L
Chlorine	<0.003 mg/L
Copper	<0.0006 mg/L in soft water;<0.03 mg/L in hard water
Gas supersaturation	<100% total gas pressure (103% for salmonid eggs/ fry)(102% for lake trout)
Hydrogen sulfide	<0.003 mg/L
Iron	<0.1 mg/L
Lead	<0.02 mg/L
Mercury	<0.0002 mg/L
Nitrate	<1.0 mg/L
Nitrite	<0.1 mg/L
Oxygen	6 mg/L for coldwater fish4 mg/L for warmwater fish
Selenium	<0.01 mg/L
Total dissolved solids	<200mg/L
Total suspended solids	<80 NTU over ambient levels
Zinc	<0.005 mg/L

In these kinds of systems fish production per unit of surface can be increased at will, as long as sufficient oxygen, fresh water and food are provided. Because of the requirement of sufficient fresh water, a massive water purification system must be integrated in the fish farm. A clever way to achieve this is the combination of hydroponic horticulture and water treatment. The exception to this rule are cages which are placed in a river or sea, which supplements the fish crop with sufficient oxygenated water. Some environmentalists object to this practice.

The cost of inputs per unit of fish weight is higher than in extensive farming, especially because of the high cost offish feed, which must contain a much higher level of protein (up to 60%) than cattle food and a balanced amino acid composition as well. However, these higher protein level requirements are a consequence of the higher food conversion efficiency (FCR—kg of feed per kg of animal produced) of aquatic animals. Fish like salmon have FCR's in the range of 1.1 kg of feed per kg of salmon whereas chickens are in the 2.5 kg of feed per kg of chicken range. Fish do not have to stand up or keep warm and this eliminates a lot of carbohydrates and fats in the diet, required to provide this energy. This frequently is offset by the lower land costs and the higher productions which can be obtained due to the high level of input control.

Figure: Expressing eggs from a female rainbow trout

Essential here is aeration of the water, as fish need a sufficient oxygen level for growth. This is achieved by bubbling, cascade flow or aqueous oxygen. Catfish, Clarias spp. can breathe atmospheric air and can tolerate much higher levels of pollutants than trout or salmon, which makes aeration and water purification less necessary and makes *Clarias* species especially suited for intensive fish production. In some *Clarias* farms about 10% of the water volume can consist of fish biomass.

The risk of infections by parasites like fish lice, fungi (Saprolegnia spp.), intestinal worms (such as nematodes or trematodes), bacteria (e.g., Yersinia spp.,Pseudomonas spp.), and protozoa (such as Dinoflagellates) is similar to animal husbandry, especially at high population densities. However, animal husbandry is a larger and more technologically mature area of human agriculture and better solutions to pathogen problem exist. Intensive aquaculture does have to provide adequate water quality (oxygen, ammonia, nitrite, etc.) levels to minimise stress, which makes the pathogen problem more difficult. This means, intensive aquaculture requires tight monitoring and a high level of expertise of the fish farmer.

Very high intensity recycle aquaculture systems (RAS), where there is control over all the production parameters, are being used for high value species. By recycling the water, very little water is used per unit of production. However, the process does have high capital and operating costs. The higher cost structures mean that RAS is only economical for high value products like broodstock for egg production, fingerlings for net pen aquaculture operations, sturgeon production, research animals and some special niche markets like live fish.

Raising ornamental cold water fish (goldfish or koi), although theoretically much more profitable due to the higher income per weight of fish produced, has never been successfully carried out until very recently. The increased incidences of dangerous viral diseases of koi Carp, together with the high value of the fish has led to initiatives in closed system koi breeding and growing in a number of countries. Today there are a few commercially successful intensive koi growing facilities in the UK, Germany and Israel.

Some producers have adapted their intensive systems in an effort to provide consumers with fish that do not carry dormant forms of viruses and diseases.

Specific Types of Fish Farms

Within intensive and extensive aquaculture methods, there are numerous specific types of fish farms; each has benefits and applications unique to its design.

Cage System

Fish cages are placed in lakes, bayous, ponds, rivers or oceans to contain and protect fish until they can be harvested. The method is also called "off-shore cultivation" when the cages are placed in the sea. They can be constructed of a wide variety of components. Fish are stocked in cages, artificially fed, and harvested when they reach market size. A few advantages of fish farming with cages are that many types of waters can be used (rivers, lakes, filled quarries, etc.), many types of fish can be raised, and fish farming can co-exist with sport fishing and other water uses. Cage farming of fishes in open seas is also gaining popularity. Concerns of disease, poaching, poor water quality, etc., lead some to believe that in general, pond systems are easier to manage and simpler to start. Also, past occurrences of cage-failures leading to escapes, have raised concern regarding the culture of non-native fish species in open-water cages. Even though the cage-industry has made numerous technological advances in cage construction in recent years, the concern for escapes remains valid.

Copper Alloys in Aquaculture: Recently, copper alloys have become important netting materials in aquaculture. Copper alloys are antimicrobial, that is, they destroy bacteria, viruses, fungi, algae, and other microbes. In the marine environment, the antimicrobial/algaecidal properties of copper alloys prevent biofouling, which can briefly be described as the undesirable accumulation, adhesion, and growth of microorganisms, plants, algae, tubeworms, barnacles, mollusks, and other organisms.

The resistance of organism growth on copper alloy nets also provides a cleaner and healthier environment for farmed fish to grow and thrive.

In addition to its antifouling benefits, copper netting has strong structural and corrosion-resistant properties in marine environments.

Copper-zinc brass alloys are currently (2011) being deployed in commercial-scale aquaculture operations in Asia, South America and the USA (Hawaii). Extensive research, including demonstrations and trials, are currently being implemented on two other copper alloys: copper-nickel and copper-silicon. Each of these alloy types has an inherent ability to reduce biofouling, cage waste, disease, and the need for antibiotics while simultaneously maintaining water circulation and oxygen requirements. Other types of copper alloys are also being considered for research and development in aquaculture operations.

Irrigation Ditch or Pond Systems

These use irrigation ditches or farm ponds to raise fish. The basic requirement is to have a ditch or pond that retains water, possibly with an above-ground irrigation system (many irrigation systems use buried pipes with headers.) Using this method, one can store one's water allotment in ponds or ditches, usually lined with bentonite clay. In small systems the fish are often fed commercial fish food, and their waste products can help fertilize the fields. In larger ponds, the pond grows water plants and algae as fish food. Some of the most successful ponds grow introduced strains of plants, as well as introduced strains of fish.

Control of water quality is crucial. Fertilizing, clarifying and pH control of the water can increase yields substantially, as long as eutrophication is prevented and oxygen levels stay high. Yields can be low if the fish grow ill from electrolyte stress.

Composite Fish Culture

The composite fish culture system is a technology developed in India by the Indian Council of Agricultural Research in the 1970s. In this system both local and imported fish species, a combination of five or six fish species is used in a single fish pond. These species are selected so that they do not compete for food among them having different types of food habitats. As a result the food available in all the parts of the pond is used. Fish used in this system include catla and silver carp which are surface feeders, rohu a column feeder and mrigal and common carp which are bottom feeders. Other fish will also feed on the excreta of the common carp and this helps contribute to the efficiency of the system which in optimal conditions will produce 3000–6000 kg of fish per hectare per year.

Integrated Recycling Systems

One of the largest problems with freshwater aquaculture is that it can use a million gallons of water per acre (about 1 m^3 of water per m^2) each year. Extended water purification systems allow for the reuse (recycling) of local water.

The largest-scale pure fish farms use a system derived (admittedly much refined) from the New Alchemy Institute in the 1970s. Basically, large plastic fish tanks are placed in a greenhouse. A hydroponic bed is placed near, above or between them. When tilapia are raised in the tanks, they are able to eat algae, which naturally grows in the tanks when the tanks are properly fertilized.

The tank water is slowly circulated to the hydroponic beds where the tilapia waste feeds commercial plant crops. Carefully cultured microorganisms in the hydroponic bed convert ammonia to nitrates, and the plants are fertilized by the nitrates and phosphates. Other wastes are strained out by the hydroponic media, which doubles as an aerated pebble-bed filter.

This system, properly tuned, produces more edible protein per unit area than any other. A wide variety of plants can grow well in the hydroponic beds. Most growers concentrate on herbs (e.g. parsley and basil), which command premium prices in small quantities all year long. The most common customers are restaurant wholesalers.

Since the system lives in a greenhouse, it adapts to almost all temperate climates, and may also adapt to tropical climates. The main environmental impact is discharge of water that must be salted to maintain the fishes' electrolyte balance. Current growers use a variety of proprietary tricks to keep fish healthy, reducing their expenses for salt and waste water discharge permits. Some veterinary authorities speculate that ultraviolet ozone disinfectant systems (widely used for ornamental fish) may play a prominent part in keeping the Tilapia healthy with recirculated water.

A number of large, well-capitalised ventures in this area have failed. Managing both the biology and markets is complicated.

Classic Fry Farming

This is also called a "Flow through system" Trout and other sport fish are often raised from eggs to fry or fingerlings and then trucked to streams and released. Normally, the fry are raised in long, shallow concrete tanks, fed with fresh stream water. The fry receive commercial fish food in pellets. While not as efficient as the New Alchemists' method, it is also far simpler, and has been used for many years to stock streams with sport fish. European eel (Anguilla anguilla) aquaculturalists procure a limited supply of glass eels, juvenile stages of the European eel which swim north from the Sargasso Sea breeding grounds, for their farms. The European eel is threatened with extinction because of the excessive catch of glass eels by Spanish fishermen and overfishing of adult eels in, e.g., the Dutch IJsselmeer, Netherlands. As per 2005, no one has managed to breed the European eel in captivity.

Issues

The issue of feeds in fish farming has been a controversial one. Many cultured fishes (tilapia, carp, catfish, many others) require no meat or fish products in their diets. Top-level carnivores (most salmon species) depend on fish feed of which a portion is usually derived from wild caught (anchovies, menhaden, etc.). Vegetable-derived proteins have successfully replaced fish meal in feeds for carnivorous fishes, but vegetable-derived oils have not successfully been incorporated into the diets of carnivores.

Secondly, farmed fish are kept in concentrations never seen in the wild (e.g. 50,000 fish in a 2-acre (8,100 m^2) area.) with each fish occupying less room than the average bathtub. This can cause several forms of pollution. Packed tightly, fish rub against each other and the sides of their cages, damaging their fins and tails and becoming sickened with various diseases and infections. This also causes stress.

However, fish tend also to be animals that aggregate into large schools at high density. Most successful aquaculture species are schooling species, which do not have social problems at high density. Aquaculturists tend to feel that operating a rearing system above its design capacity or above the social density limit of the fish will result in decreased growth rate and increased FCR (food conversion ratio - kg dry feed/kg of fish produced), which will result in increased cost and risk of health problems along with a decrease in profits. Stressing the animals is not desirable, but the concept of and measurement of stress must be viewed from the perspective of the animal using the scientific method.

Sea lice, particularly *Lepeophtheirus salmonis* and various *Caligus* species, including *Caligus clemensi* and *Caligus rogercresseyi*, can cause deadly infestations of both farm-grown and wild salmon. Sea lice are ectoparasites which feed on mucus, blood, and skin, and migrate and latch onto the skin of wild salmon during free-swimming, planktonic *nauplii* and *copepodid* larval stages, which can persist for several days. Large numbers of highly populated, open-net salmon farms can create exceptionally large concentrations of sea lice; when exposed in river estuaries containing large numbers of open-net farms, many young wild salmon are infected, and do not survive as a result. Adult salmon may survive otherwise critical numbers of sea lice, but small, thin-skinned juvenile salmon migrating to sea are highly vulnerable. On the Pacific coast of Canada, the louse-induced mortality of pink salmon in some regions is commonly over 80%.

A 2008 meta-analysis of available data shows that salmon farming reduces the survival of associated wild salmon populations. This relationship has been shown to hold for Atlantic, steelhead, pink, chum, and coho salmon. The decrease in survival or abundance often exceeds 50 percent.

Diseases and parasites are the most commonly cited reasons for such decreases. Some species of sea lice have been noted to target farmed coho and Atlantic salmon. Such parasites have been shown to have an effect on nearby wild fish. One place that has garnered international media attention is British Columbia's Broughton Archipelago. There, juvenile wild salmon must "run a gauntlet" of large fish farms located off-shore near river outlets before making their way to sea. It is alleged that the farms cause such severe sea lice infestations that one study predicted in 2007 a 99% collapse in the wild salmon population by 2011. This claim, however, has been criticized by numerous scientists who question the correlation between increased fish farming and increases in sea lice infestation among wild salmon.

Because of parasite problems, some aquaculture operators frequently use strong antibiotic drugs to keep the fish alive (but many fish still die prematurely at rates of up to 30 percent). In some cases, these drugs have entered the environment. Additionally, the residual presence of these drugs in human food products has become controversial. Use of antibiotics in food production is thought to increase the prevalence of antibiotic resistance in human diseases. At some facilities, the use of antibiotic drugs in aquaculture has decreased considerably due to vaccinations and other techniques. However, most fish farming operations still use antibiotics, many of which escape into the surrounding environment.

The lice and pathogen problems of the 1990s facilitated the development of current treatment methods for sea lice and pathogens. These developments reduced the stress from parasite/pathogen problems. However, being in an ocean environment, the transfer of disease organisms from the wild fish to the aquaculture fish is an ever-present risk.

The very large number of fish kept long-term in a single location contributes to habitat destruction of the nearby areas. The high concentrations of fish produce a significant amount of condensed faeces, often contaminated with drugs, which again affect local waterways. However, these effects are very local to the actual fish farm site and are minimal to non-measurable in high current sites.

Concern remains that resultant bacterial growth strips the water of oxygen, reducing or killing off the local marine life. Once an area has been so contaminated, the fish farms are moved to new, uncontaminated areas. This practice has angered nearby fishermen.

Other potential problems faced by aquaculturists are the obtaining of various permits and water-use rights, profitability, concerns about invasive species and genetic engineering depending on what species are involved, and interaction with the United Nations Convention on the Law of the Sea.

In regards to genetically modified farmed salmon, concern has been raised over their proven reproductive advantage and how it could potentially decimate local fish populations, if released into the wild. Biologist Rick Howard did a controlled laboratory study where wild fish and GMO fish were allowed to breed. The GMO fish crowded out the wild fish in spawning beds, but the offspring were less likely to survive. The colourant used to make pen-raised salmon appear rosy like their wild cousins has been linked with retinal problems in humans.

Labelling

In 2005, Alaska passed legislation requiring that any genetically altered fish sold in the state be labelled. In 2006, a Consumer Reports investigation revealed that farm-raised salmon is frequently sold as wild.

In 2008, the US National Organic Standards Board allowed farmed fish to be labelled as organic provided less than 25% of their feed came from wild fish. This decision was criticized by the advocacy group Food & Water Watch as "bending the rules" about organic labelling. In the European Union, fish labelling as to species, method of production and origin, has been required since 2002.

Concerns continue over the labelling of salmon as farmed or wild caught, as well as about the humane treatment of farmed fish. The Marine Stewardship Council has established an Eco label to distinguish between farmed and wild caught salmon, while the RSPCA has established the Freedom Food label to indicate humane treatment of farmed salmon as well as other food products.

Indoor Fish Farming

An alternative to outdoor open ocean cage aquaculture, is through the use of a recirculation aquaculture system (RAS). A RAS is a series of culture tanks and filters where water is continuously recycled and monitored to keep optimal conditions year round. To prevent the deterioration of water quality, the water is treated mechanically through the removal of particulate matter and biologically through the conversion of harmful accumulated chemicals into nontoxic ones.

Other treatments such as UV sterilization, ozonation, and oxygen injection are also used to maintain optimal water quality. Through this system, many of the environmental drawbacks of aquaculture are minimised including escaped fish, water usage, and the introduction of pollutants. The practices also increased feed-use efficiency growth by providing optimum water quality (Timmons et al., 2002; Piedrahita, 2003).

One of the drawbacks to recirculation aquaculture systems is water exchange. However, the rate of water exchange can be reduced through aquaponics, such as the incorporation of hydroponically grown plants (Corpron and Armstrong, 1983) and denitrification (Klas et al., 2006). Both methods reduce the amount of nitrate in the water, and can potentially eliminate the need for water exchanges, closing the aquaculture system from the environment. The amount of interaction between the aquaculture system and the environment can be measured through the cumulative feed burden (CFB kg/M3), which measures the amount of feed that goes into the RAS relative to the amount of water and waste discharged.

Because of its high capital and operating costs, RAS has generally been restricted to practices such as broodstock maturation, larval rearing, fingerling production, research animal production, SPF (specific pathogen free) animal production, and caviar and ornamental fish production. Although the use of RAS for other species is considered by many aquaculturalists to be impractical, there has been some limited successful implementation of this with high value product such as barramundi, sturgeon and live tilapia in the US.

Slaughter Methods

Tanks saturated with carbon dioxide have been used to make fish unconscious. Then their gills are cut with a knife so that the fish bleed out before they are further processed. This is no longer considered a humane method of slaughter. Methods that induce much less physiological stress are electrical or percussive stunning and this has led to the phasing out of the carbon dioxide slaughter method in Europe.

Inhumane Methods

According to T. Håstein of the National Veterinary Institute, "Different methods for slaughter of fish are in place and it is no doubt that many of them may be considered as appalling from an animal welfare point of view." A 2004 report by the EFSA Scientific Panel on Animal Health and Welfare explained: "Many existing commercial killing methods expose fish to substantial suffering over a prolonged period of time. For some species, existing methods, whilst capable of killing fish humanely, are not doing so because operators don't have the knowledge to evaluate them." Following are some of the less humane ways of killing fish.

- *Air Asphyxiation.* This amounts to suffocation in the open air. The process can take upwards of 15 minutes to induce death, although unconsciousness typically sets in sooner.
- *Ice baths / chilling.* Farmed fish are sometimes chilled on ice or submerged in near-freezing water. The purpose is to dampen muscle movements by the fish and to delay the onset of post-death decay. However, it does not necessarily reduce sensibility to pain; indeed, the chilling process has been shown to elevate cortisol. In addition, reduced body temperature extends the time before fish lose consciousness.
- *CO_2 narcosis.*
- *Exsanguination without stunning.* This is a process in which fish are taken up from water, held still, and cut so as to cause bleeding. According to references in Yue , this can leave fish writhing for an average of four minutes, and some catfish still responded to noxious stimuli after more than 15 minutes.

More Humane Methods

- *Percussive stunning.*
- *Electric stunning.* This can be humane when a proper current, duration, conductivity, and temperature are present. One advantage is that in-water

stunning allows fish to be rendered unconscious without stressful handling or displacement. However, improper stunning may not induce insensibility long enough to prevent the fish from enduring exsanguination while conscious. It's unknown whether the optimal stunning parameters that researchers have determined in studies are used by the industry in practice.

Aquaculture of Catfish

Catfish are easy to farm in warm climates, leading to inexpensive and safe food at local grocers. Catfish raised in inland tanks or channels are considered safe for the environment, since their waste and disease should be contained and not spread to the wild.

Asia

In Asia, many catfish species are important as food. Several walking catfish (Clariidae) and shark catfish (Pangasiidae) species are heavily cultured in Africa and Asia. Exports of one particular shark catfish species from Vietnam, *Pangasius bocourti*, has met with pressures from the U.S. catfish industry. In 2003, The United States Congress passed a law preventing the imported fish from being labelled as catfish. As a result, the Vietnamese exporters of this fish now label their products sold in the US as "basa fish."

United States

Ictalurids are cultivated in North America (especially in the Deep South, with Mississippi being the largest domestic catfish producer). Channel catfish (*Ictalurus punctatus*) supports a $450 million/yr aquaculture industry. The US farm-raised catfish industry began in the early 1960s in Kansas, Oklahoma and Arkansas.

Channel catfish quickly became the major catfish grown, as it was hardy and easily spawned in earthen ponds. By the late 60s, the industry moved into the Mississippi Delta as farmers struggled with sagging profits in cotton, rice and soybeans, especially on those farm areas where soils had a very high clay content.

The Mississippi Deltaic Plain includes two active pro-grading deltas: the modern bird-foot Balize delta, commonly referred to as the Mississippi Delta, and the Atchafalya delta.

In addition, there are degrading deltaic systems, such as the Lafourche and the St. Bernard (ref: World Delta Data Base, Hart and Coleman.) These deltas became the industry home for the catfish industry, as they had the soils, climate and shallow aquifersto provide water for the earthen ponds that grow 360-380 million pounds (160,000 to 170,000 tons) of catfish annually. Catfish are fed a grain-based diet that includes soybean meal. Fish are fed daily through the summer at rates of 1-6% of body weight with the pelleted floating feed. Catfish need about two pounds of feed to produce one pound of live weight. Mississippi is home to 100,000 acres (400 km^2) of catfish ponds, the largest of any state. Other states important in growing catfish include Alabama and Louisiana.

Aquarium

There is a large and growing ornamental fish trade, with hundreds of species of catfish, such as *Corydoras* and armoured suckermouth catfish (often called plecos), being a popular component of many aquaria. Other catfish commonly found in the aquarium trade are banjo catfish, talking catfish, and long-whiskered catfish.

2

Characteristics of Ornamental Fish Farming

Channel Catfish

Channel catfish, *Ictalurus punctatus*, is North America's most numerous catfish species. It is the official fish of Missouri, Iowa, Nebraska, Kansas, and Tennessee, and is informally referred to as a "channel cat". In the United States they are the most fished catfish species with approximately 8 million anglers targeting them per year. The popularity of channel catfish for food has contributed to the rapid growth of aquaculture of this species in the United States.

Distribution and Habitat

Channel catfish are native to the Nearctic, being well distributed in lower Canada and the eastern and northern United States, as well as parts of northern Mexico. They have also been introduced into some waters of landlocked Europe and parts of Malaysia and almost many parts of Indonesia. They thrive in small and large rivers, reservoirs, natural lakes, and ponds. Channel "cats" are cavity nesters, meaning they lay their eggs in crevices, hollows, or debris, in order to protect them from swift currents. In Canada, the species is largely, though not exclusively, limited to the Great Lakes watershed from Lake Nipigon southward.

Characteristics

Channel catfish possess very keen senses of smell and taste. At the pits of their nostrils (nares) are very sensitive odour sensing organs with a very high concentration of olfactory receptors. In channel catfish these organs are sensitive enough to detect several amino acids at about 1 part per 100 million in water. In addition, channel catfish have taste buds distributed over the surface of their entire body. These buds are especially concentrated on the channel catfish's four pair of barbels (whiskers) surrounding the mouth — about 25 buds per square millimetre. This combination of exceptional senses of taste and smell allows the channel catfish to find food in dark, stained, or muddy water with relative ease.

Figure: Chuck the Channel catfish, 1986 roadside sculpture in Selkirk, ManitobaCanada

Fishing

Channel catfish are omnivores and can be caught using a variety of natural and prepared baits, including crickets, night crawlers, minnows, shad, crawfish, frogs, bullheads, sunfish, and suckers. Catfish have even been known to take Ivory Soap as bait . Another method of catching catfish is using stinkbaits, which are prepared with dead fish, crawfish, garlic, blood, liver, meat, cheese, dough, and even Kool-Aid powder. Sometimes these stinkbaits are prepared into a doughball and mashed onto a hook; other times they are smeared in special tubes meant to hold these baits, and fished slowly on the bottom. Grocery store baits such as chicken livers, shrimp, dog food, squid, and bubble gum will also catch plenty of channel cats.

Juglines, trotlines, limb lines and bank lines are popular methods of fishing for channel catfish in addition to traditional rod and reel fishing. Another method uses traps, either "slat traps" — long wooden traps with an angled entrance — and wire hoop traps. Typical bait for these traps include rotten cheese and dog food. Catches of as many as 100+ fish a day are common in catfish traps.

When removing the hook from a catfish, anglers should be mindful of the sharp spines on the pectoral and dorsal fin.

Length and Weight

A member of the *Ictalurus* genus of American catfishes, channel catfish have a top-end size of approximately 40-50 pounds (18–23 kg). The world record channel catfish weighed 58 pounds and was taken from the Santee-Cooper Reservoir in South Carolina, July 7, 1964. Realistically, a channel catfish over 20 pounds (9 kg) is a spectacular specimen, and most catfish anglers view a 10 pound (4.5 kg) fish as a very admirable catch. Furthermore the average size channel catfish an angler could expect to find in most waterways would be between 2 and 4 pounds.

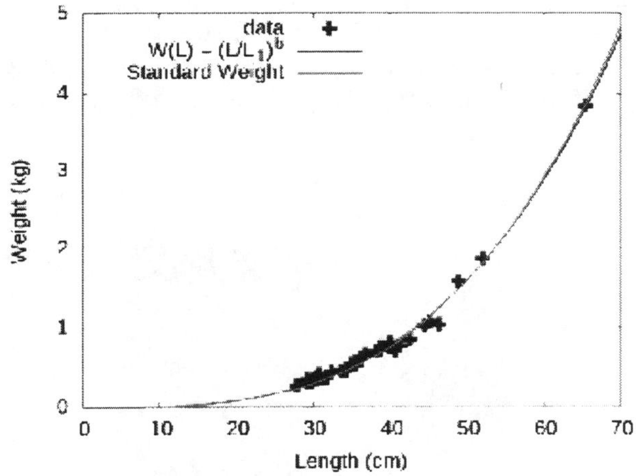

Figure: Weight vs. length for Channel Catfish, where b = 3.2293 and $L_1 = 45.23$ cm.

Channel catfish will often coexist in the same waterways with its close relatives, blue catfish, which are somewhat less common but tend to grow a lot larger (with several specimen confirmed to weight above the 100 lb. mark).

As channel catfish grow longer, they increase in weight. The relationship between length and weight is not linear, but instead follows a pareto law. The relationship between length (L, in cm) and weight (W, in kg) for nearly all species of fish can be expressed by an equation of the form:

$$W = (L / L_1)^b$$

Invariably, b is close to 3.0 for all species, L_1 is the length of a typical fish weighing 1 kg. For channel catfish, b = 3.2293, somewhat higher than for many common species, and $L_1 = 45.23$ cm.

Aquaculture

Aquaculture, also known as aquafarming, is the farming of aquatic organisms such as fish, crustaceans, molluscs and aquatic plants. Aquaculture involves cultivating freshwater and saltwater populations under controlled conditions, and can be contrasted with commercial fishing, which is the harvesting of wild fish. Mariculture refers to aquaculture practiced in marine environments.

The reported output from global aquaculture operations would supply one half of the fish and shellfish that is directly consumed by humans; however, there are issues about the reliability of the reported figures. Further, in current aquaculture practice, products from several pounds of wild fish are used to produce one pound of a piscivorous fish like salmon.

Particular kinds of aquaculture include fish farming, shrimp farming, oyster farming, algaculture (such as seaweed farming), and the cultivation of ornamental fish. Particular methods include aquaponics and Integrated multi-trophic aquaculture, both of which integrate fish farming and plant farming.

History

Figure: Workers harvest catfish from the Delta Pride Catfish farms in Mississippi

The indigenous Gunditjmara people in Victoria, Australia may have raised eels as early as 6000 BC. There is evidence that they developed about 100 square kilometres (39 sq mi) of volcanic floodplains in the vicinity of Lake Condah into a complex of channels and dams, that they used woven traps to capture eels, and that capturing and smoking eels supported them year round.

Aquaculture was operating in China circa 2500 BC. When the waters subsided after river floods, some fishes, mainlycarp, were trapped in lakes. Early aquaculturists fed their brood using nymphs and silkworm feces, and ate them.

A fortunate genetic mutation of carp led to the emergence of goldfish during the Tang Dynasty. Japanese cultivated seaweed by providing bamboo poles and, later, nets and oyster shells to serve as anchoring surfaces for spores.

Romans Bred Fish in Ponds: In central Europe, early Christian monasteries adopted Roman aquacultural practices. Aquaculture spread in Europe during the Middle Ages, since away from the seacoasts and the big rivers, fish were scarce/ expensive. Improvements in transportation during the 19th century made fish easily available and inexpensive, even in inland areas, making aquaculture less popular.

Hawaiians constructed oceanic fish ponds. A remarkable example is a fish pond dating from at least 1,000 years ago, at Alekoko. Legend says that it was constructed by the mythical Menehune dwarf people.

In 1859 Stephen Ainsworth of West Bloomfield, New York, began experiments with brook trout. By 1864 Seth Green had established a commercial fish hatching operation at Caledonia Springs, near Rochester, New York. By 1866, with the involvement of Dr. W. W. Fletcher of Concord, Massachusetts, artificial fish hatcheries were under way in both Canada and the United States.

When the Dildo Island fish hatchery opened in Newfoundland in 1889, it was the largest and most advanced in the world. Californians harvested wild kelp and attempted to manage supply circa 1900, later labelling it a wartime resource.

According to the FAO, aquaculture "is understood to mean the farming of aquatic organisms including fish, molluscs, crustaceans and aquatic plants. Farming implies some form of intervention in the rearing process to enhance production, such as regular stocking, feeding, protection from predators, etc. Farming also implies individual or corporate ownership of the stock being cultivated."

1st Century Practice

About 430 (97%) of the species cultured as of 2007 were domesticated during the 20th century, of which an estimated 106 came in the decade to 2007. Given the long-term importance of agriculture, it is interesting to note that to date only 0.08% of known land plant species and 0.0002% of known land animal species have been domesticated, compared with 0.17% of known marine plant species and 0.13% of known marine animal species.

Domestication typically involves about a decade of scientific research. Domesticating aquatic species involves fewer risks to humans than land animals, which took a large toll in human lives. Most major human diseases originated in domesticated animals, through diseases such as smallpox and diphtheria, that like most infectious diseases, move to humans from animals. No human pathogens of comparable virulence have yet emerged from marine species.

Definition

Figure: Tilapia, a commonly farmed fish due to its adaptability

Harvest stagnation in wild fisheries and overexploitation of popular marine species, combined with a growing demand for high quality protein, encourage aquaculturists to domesticate other marine species.

Production Volume

Table: Top ten species groups in 2004

Species Group	*Million Tonnes*
Carps and other cyprinids	18.30
Oysters	4.60
Clams, cockles, ark shells	4.12
Miscellaneous freshwater fishes	3.74
Shrimps, prawns	2.48
Salmons, trouts, smelts	1.98
Mussels	1.86
Tilapias and other cichlids	1.82
Scallops, pectens	1.17
Miscellaneous marine molluscs	1.07

In 2004, the total world production of fisheries was 140 million tonnes of which aquaculture contributed 45 million tonnes, about one third. The growth rate of worldwide aquaculture has been sustained and rapid, averaging about 8 percent per annum for over thirty years, while the take from wild fisheries has been essentially flat for the last decade. The aquaculture market reached $86 billion in Aquaculture is an especially important economic activity in China. Between 1980 and 1997, the Chinese Bureau of Fisheries reports, aquaculture harvests grew at an annual rate of 16.7 percent, jumping from 1.9 million tonnes to nearly 23 million tonnes. In 2005, China accounted for 70% of world production. Aquaculture is also currently one of the fastest growing areas of food production in the U.S.

Table: Top ten aquaculture producers in 2004

Country	***Million Tonnes***
China	30.61
India	2.47
Vietnam	1.20
Thailand	1.17
Indonesia	1.05
Bangladesh	0.91
Japan	0.78
Chile	0.67
Norway	0.64
United States	0.61
Other countries	5.35
Total	45.47

Approximately 90% of all U.S. shrimp consumption is farmed and imported. In recent years salmon aquaculture has become a major export in southern Chile, especially in Puerto Montt, Chile's fastest-growing city.

Over Reporting

China overwhelmingly dominates the world in reported aquaculture output. They report a total output which is double that of the rest of the world put together. However, there are issues with the accuracy of China's returns.

In 2001, the fisheries scientists Reg Watson and Daniel Pauly expressed concerns in a letter to *Nature*, that China was over reporting its catch from wild fisheries in the 1990s. They said that made it appear that the global catch since 1988 was increasing annually by 300,000 tonnes, whereas it was really shrinking annually by 350,000 tonnes. Watson and Pauly suggested this may be related to China policies where state entities that monitor the economy are also tasked with increasing output. Also, until recently, the promotion of Chinese officials was based on production increases from their own areas.

China disputes this claim. The official Xinhua News Agency quoted Yang Jian, director general of the Agriculture Ministry's Bureau of Fisheries, as saying that China's figures were "basically correct". However, the FAO accepts there are issues with the reliability of China's statistical returns, and currently treats data from China, including the aquaculture data, apart from the rest of the world.

Methods

Mariculture

Mariculture is the term used for the cultivation of marine organisms in seawater, usually in sheltered coastal waters. In particular, the farming of marine fish is an example of mariculture, and so also is the farming of marine crustaceans (such as shrimps), molluscs (such as oysters) and seaweed.

Integrated

Integrated Multi-Trophic Aquaculture (IMTA) is a practice in which the by-products (wastes) from one species are recycled to become inputs (fertilizers, food) for another. Fed aquaculture (for example, fish, shrimp) is combined with inorganic extractive (for example, seaweed) and organic extractive (for example, shellfish) aquaculture to create balanced systems for environmental sustainability (biomitigation), economic stability (product diversification and risk reduction) and social acceptability (better management practices).

"Multi-Trophic" refers to the incorporation of species from different trophic or nutritional levels in the same system. This is one potential distinction from the age-old practice of aquatic polyculture, which could simply be the co-culture of different fish species from the same trophic level.

In this case, these organisms may all share the same biological and chemical processes, with few synergistic benefits, which could potentially lead to significant shifts in the ecosystem. Some traditional polyculture systems may, in fact, incorporate a greater diversity of species, occupying several niches, as extensive cultures (low intensity, low management) within the same pond. The "Integrated" in IMTA refers to the more intensive cultivation of the different species in proximity of each other, connected by nutrient and energy transfer through water.

Ideally, the biological and chemical processes in an IMTA system should balance. This is achieved through the appropriate selection and proportions of different species providing different ecosystem functions. The co-cultured species are typically more than just biofilters; they are harvestable crops of commercial value. A working IMTA system can result in greater total production based on mutual benefits to the co-cultured species and improved ecosystem health, even if the production of individual species is lower than in a monoculture over a short term period.

Sometimes the term "Integrated Aquaculture" is used to describe the integration of monocultures through water transfer. For all intents and purposes however, the terms "IMTA" and "integrated aquaculture" differ only in their degree of descriptiveness. Aquaponics, fractionated aquaculture, IAAS (integrated agriculture-aquaculture systems), IPUAS (integrated peri-urban-aquaculture systems), and IFAS (integrated fisheries-aquaculture systems) are other variations of the IMTA concept.

Netting Materials

Various materials, including nylon, polyester, polypropylene, polyethylene, plastic-coated welded wire, rubber, patented rope products (Spectra, Thorn-D, Dyneema), galvanised steel and copper are used for netting in aquaculture fish enclosures around the world. All of these materials are selected for a variety of reasons, including design feasibility, material strength, cost, and corrosion resistance.

Copper Alloys in Aquaculture: Recently, copper alloys have become important netting materials in aquaculture because they are antimicrobial (i.e., they destroy bacteria, viruses, fungi, algae, and other microbes) and they therefore prevent biofouling (i.e., the undesirable accumulation, adhesion, and growth of microorganisms, plants, algae, tubeworms, barnacles, mollusks, and other organisms). By inhibiting microbial growth, copper alloy aquaculture cages avoid costly net changes that are necessary with other materials. The resistance of organism growth on copper alloy nets also provides a cleaner and healthier environment for farmed fish to grow and thrive.

Species Groups

Fish

The farming of fish is the most common form of aquaculture. It involves raising fish commercially in tanks, ponds, or ocean enclosures, usually for food. A facility that releases juvenile fish into the wild for recreational fishing or to supplement a species' natural numbers is generally referred to as a fish hatchery. Fish species raised by fish farms include salmon, bigeye tuna, carp, tilapia, catfish and cod.

In the Merranean, young bluefin tuna are netted at sea and towed slowly towards the shore. They are then interned in offshore pens where they are further grown for the market. In 2009, researchers in Australia managed for the first time to coax tuna (Southern bluefin) to breed in landlocked tanks.

Crustaceans

Commercial shrimp farming began in the 1970s, and production grew steeply thereafter. Global production reached more than 1.6 million tonnes in 2003, worth about 9 billion U.S. dollars. About 75% of farmed shrimp is produced in Asia, in particular in China and Thailand. The other 25% is produced mainly in Latin America, where Brazil is the largest producer. Thailand is the largest exporter.

Shrimp farming has changed from its traditional, small-scale form in Southeast Asia into a global industry. Technological advances have led to ever higher densities per unit area, and broodstock is shipped worldwide. Virtually all farmed shrimp are penaeids (i.e., shrimp of the family *Penaeidae*), and just two species of shrimp, the Pacific white shrimp and the giant tiger prawn, account for about 80% of all farmed shrimp. These industrial monocultures are very susceptible to disease, which has decimated shrimp populations across entire regions. Increasing ecological problems, repeated disease outbreaks, and pressure and criticism from both NGOs and consumer countries led to changes in the industry in the late 1990s and generally stronger regulations. In 1999, governments, industry representatives, and environmental organisations initiated a program aimed at developing and promoting more sustainable farming practices.

Freshwater prawn farming shares many characteristics with, including many problems with marine shrimp farming. Unique problems are introduced by the developmental life cycle of the main species, the giant river prawn.

The global annual production of freshwater prawns (excluding crayfish and crabs) in 2003 was about 280,000 tonnes of which China produced 180,000 tonnes followed by India and Thailand with 35,000 tonnes each. Additionally, China produced about 370,000 tonnes of Chinese river crab.

Molluscs

Abalone farming began in the late 1950s and early 1960s in Japan and China. Since the mid-1990s, this industry has become increasingly successful. Over-fishing and poaching have reduced wild populations to the extent that farmed abalone now supplies most abalone meat.

Echinoderms

Commercially harvested echinoderms include sea cucumbers and sea urchins. In China, sea cucumbers are farmed in artificial ponds as large as 1,000 acres (400 ha).

Algae

Microalgae, also referred to as phytoplankton, microphytes, or planktonic algae constitute the majority of cultivated algae. Macroalgae, commonly known as seaweed, also have many commercial and industrial uses, but due to their size and specific requirements, they are not easily cultivated on a large scale and are most often taken in the wild.

Issues

Aquaculture can be more environmentally damaging than exploiting wild fisheries on a local area basis but has considerably less impact on the global environment on a per kg of production basis. Local concerns include waste handling, side-effects of antibiotics, competition between farmed and wild animals, and using other fish to feed more marketable carnivorous fish. However, research and commercial feed improvements during the 1990s & 2000s have lessened many of these.

Fish waste is organic and composed of nutrients necessary in all components of aquatic food webs. In-ocean aquaculture often produces much higher than normal fish waste concentrations. The waste collects on the ocean bottom, damaging or eliminating bottom-dwelling life. Waste can also decrease dissolved oxygen levels in the water column, putting further pressure on wild animals.

Fish Oils

Tilapia from aquaculture has been shown to contain more fat and a much higher ratio of omega-6 to omega-3 oils.

Impacts on Wild Fish

Salmon farming currently leads to a high demand for wild forage fish. Fish do not actually produce omega-3 fatty acids, but instead accumulate them from either consuming microalgae that produce these fatty acids, as is the case with forage fish like herring and sardines, or, as is the case with fatty predatory fish, like salmon, by eating prey fish that have accumulated omega-3 fatty acids from microalgae. To satisfy this requirement, more than 50 percent of the world fish oil production is fed to farmed salmon.

In addition, as carnivores, salmon require large nutritional intakes of protein, protein which is often supplied to them in the form of forage fish. Consequently, farmed salmon consume more wild fish than they generate as a final product. To produce one pound of farmed salmon, products from several pounds of wild fish are fed to them. As the salmon farming industry expands, it requires more wild forage fish for feed, at a time when seventy five percent of the worlds monitored fisheries are already near to or have exceeded their maximum sustainable yield. The industrial scale extraction of wild forage fish for salmon farming then impacts the survivability of the wild predator fish who rely on them for food.

Fish can escape from coastal pens, where they can interbreed with their wild counterparts, diluting wild genetic stocks. Escaped fish can become invasive, out competing native species.

Coastal Ecosystems

Aquaculture is becoming a significant threat to coastal ecosystems. About 20 percent of mangrove forests have been destroyed since 1980, partly due to shrimp farming. An extended cost–benefit analysis of the total economic value of shrimp aquaculture built on mangrove ecosystems found that the external costs were much higher than the external benefits. Over four decades, 269,000 hectares (660,000 acres) of Indonesian mangroves have been converted to shrimp farms. Most of these farms are abandoned within a decade because of the toxin build-up and nutrient loss.

Salmon farms are typically sited in pristine coastal ecosystems which they then pollute. A farm with 200,000 salmon discharges more fecal waste than a city of 60,000 people. This waste is discharged directly into the surrounding aquatic environment, untreated, often containing antibiotics and pesticides." There is also an accumulation of heavy metals on the benthos (seafloor) near the salmon farms, particularly copper andzinc.

Genetic Modification

Salmon have been genetically modified for faster growth, although they are not approved for commercial use, in the face of opposition. One study, in a laboratory setting, found that modified salmon mixed with their wild relatives were aggressive in competing, but ultimately failed.

Animal Welfare

As with the farming of terrestrial animals, social attitudes influence the need for humane practices and regulations in farmed marine animals. Under the guidelines advised by the Farm Animal Welfare Council good animal welfare means both fitness and a sense of well being in the animal's physical and mental state. This can be defined by the Five Freedoms:

- Freedom from hunger & thirst
- Freedom from discomfort
- Freedom from pain, disease, or injury
- Freedom to express normal behaviour
- Freedom from fear and distress

However, the controversial issue in aquaculture is whether fish and farmed marine invertebrates are actually sentient, or have the perception and awareness to experience suffering. Although no evidence of this has been found in marine invertebrates, recent studies conclude that fish do have the necessary receptors (nociceptors) to sense noxious stimuli and so are likely to experience states of pain, fear and stress. Consequently, welfare in aquaculture is directed at vertebrates; finfish in particular.

Common Welfare Concerns

Welfare in aquaculture can be impacted by a number of issues such as stocking densities, behavioural interactions, disease and parasitism. A major problem in determining the cause of impaired welfare is that these issues are often all interrelated and influence each other at different times.

Optimal stocking density is often defined by the carrying capacity of the stocked environment and the amount of individual space needed by the fish, which is very species specific. Although behavioural interactions such as shoaling may mean that high stocking densities are beneficial to some species, in many cultured species high stocking densities may be of concern. Crowding can constrain normal swimming behaviour, as well as increase aggressive and competitive behaviours such as cannibalism, feed competition, territoriality and dominance/subordination hierarchies. This potentially increases the risk of tissue damage due to abrasion from fish-to-fish contact or fish-to-cage contact. Fish can suffer reductions in food intake and food conversion efficiency. In addition, high stocking densities can result in water flow being insufficient, creating inadequate oxygen supply and waste product removal. Dissolved oxygen is essential for fish respiration and concentrations below critical levels can induce stress and even lead to asphyxiation. Ammonia, a nitrogen excretion product, is highly toxic to fish at accumulated levels, particularly when oxygen concentrations are low.

Many of these interactions and effects cause stress in the fish, which can be a major factor in facilitating fish disease. For many parasites, infestation depends on the host's degree of mobility, the density of the host population and vulnerability of the host's defence system. Sea lice are the primary parasitic problem for finfish in aquaculture, high numbers causing widespread skin erosion and haemorrhaging, gill congestion, and increased mucus production. There are also a number of prominent viral and bacterial pathogens that can have severe effects on internal organs and nervous systems.

Improving Welfare

The key to improving welfare of marine cultured organisms is to reduce stress to a minimum, as prolonged or repeated stress can cause a range of adverse effects. Attempts to minimise stress can occur throughout the culture process. During grow out it is important to keep stocking densities at appropriate levels specific to each species, as well as separating size classes and grading to reduce aggressive behavioural interactions. Keeping nets and cages clean can assist positive water flow to reduce the risk of water degradation.

Not surprisingly disease and parasitism can have a major effect on fish welfare and it is important for farmers not only to manage infected stock but also to apply disease prevention measures. However, prevention methods, such as vaccination, can also induce stress because of the extra handling and injection. Other methods include adding antibiotics to feed, adding chemicals into water for treatment baths and biological control, such as using cleaner wrasse to remove lice from farmed salmon.

Many steps are involved in transport, including capture, food deprivation to reduce faecal contamination of transport water, transfer to transport vehicle via nets or pumps, plus transport and transfer to the delivery location. During transport water needs to be maintained to a high quality, with regulated temperature, sufficient oxygen and minimal waste products. In some cases anaesthetics may be used in small doses to calm fish before transport.

Aquaculture is sometimes part of an environmental rehabilitation program or as an aid in conserving endangered species.

Prospects

Global wild fisheries are in decline, with valuable habitat such as estuaries in critical condition. The aquaculture or farming of piscivorous fish, like salmon, does not help the problem because they need to eat products from other fish, such as fish meal and fish oil. Studies have shown that salmon farming has major negative impacts on wild salmon, as well as the forage fish that need to be caught to feed them. Fish that are higher on the food chain are less efficient sources of food energy.

Apart from fish and shrimp, some aquaculture undertakings, such as seaweed and filter-feeding bivalve mollusks like oysters, clams, mussels and scallops, are relatively benign and even environmentally restorative. Filter-feeders filter pollutants as well as nutrients from the water, improving water quality. Seaweeds extract nutrients such as inorganic nitrogen and phosphorus directly from the water, and filter-feeding mollluskscan extract nutrients as they feed on particulates, such as phytoplankton and detritus.

Some profitable aquaculture cooperatives promote sustainable practices. New methods lessen the risk of biological and chemical pollution through minimising fish stress, fallowing netpens, and applying Integrated Pest Management. Vaccines are being used more and more to reduce antibiotic use for disease control.

Onshore recirculating aquaculture systems, facilities using polyculture techniques, and properly sited facilities (for example, offshore areas with strong currents) are examples of ways to manage negative environmental effects.

Recirculating aquaculture systems (RAS) recycle water by circulating it through filters to remove fish waste and food and then recirculating it back into the tanks. This saves water and the waste gathered can be used incompost or, in some cases, could even be treated and used on land. While RAS was developed with freshwater fish in mind, scientist associated with the Agricultural Research Service have found a way to rear saltwater fish using RAS in low-salinity waters. Although saltwater fish are raised in off-shore cages or caught with nets in water that typically has a salinity of 35 parts per thousand (ppt), scientists were able to produce healthy pompano, a saltwater fish, in tanks with a salinity of only 5 ppt. Commercialising low-salinity RAS are predicted to have positive environmental and economical effects.

Unwanted nutrients from the fish food would not be added to the ocean and the risk of transmitting diseases between wild and farm-raised fish would greatly be reduced. The price of expensive saltwater fish, such as the pompano and combia used in the experiments, would be reduced. However, before any of this can be done researchers must study every aspect of the fish's lifecycle, including the amount of ammonia and nitrate the fish will tolerate in the water, what to feed the fish during each stage of its lifecycle, the stocking rate that will produce the healthiest fish, etc.

Aquaponics

Figure: A small, portable aquaponics system

Aquaponics is a sustainable food production system that combines a traditional aquaculture (raising aquatic animals such as snails, fish, crayfish or prawns in tanks) with hydroponics (cultivating plants in water) in a symbiotic environment. In the aquaculture, effluents accumulate in the water, increasing toxicity for the fish. This water is led to a hydroponic system where the by-products from the aquaculture are filtered out by the plants as vital nutrients, after which the cleansed water is recirculated back to the animals. The term *aquaponics* is a portmanteau of the terms *aquaculture* and *hydroponic.*

Aquaponic systems vary in size from small indoor or outdoor units to large commercial units, using the same technology. The systems usually contain fresh water, but salt water systems are plausible depending on the type of aquatic animal and which plants. Aquaponic science may still be considered to be at an early stage.

Function

Aquaponics consists of two main parts, with the aquaculture part for raising aquatic animals and the hydroponics part for growing plants. Aquatic effluents resulting from uneaten feed or raising animals like fish, accumulates in water due to the closed system recirculation of most aquaculture systems. The effluent-rich water becomes toxic to the aquatic animal in high concentrations but these effluents are nutrients essential for plant growth. Although consisting primarily of these two parts, aquaponics systems are usually grouped into several components or subsystems responsible for the effective removal of solid wastes, for adding bases to neutralise acids, or for maintaining water oxygenation. Typical components include:

- *Rearing tank*: the tanks for raising and feeding the fish;
- *Solids removal*: a unit for catching uneaten food and detached biofilms, and for settling out fine particulates;
- *Biofilter*: a place where the nitrification bacteria can grow and convert ammonia into nitrates, which are usable by the plants;
- *Hydroponics subsystem*: the portion of the system where plants are grown by absorbing excess nutrients from the water;
- *Sump*: the lowest point in the system where the water flows to and from which it is pumped back to the rearing tanks.

Depending on the sophistication and cost of the aquaponics system, the units for solids removal, biofiltration, and/or the hydroponics subsystem may be combined into one unit or subsystem, which prevents the water from flowing directly from the aquaculture part of the system to the hydroponics part.

Nitrification

Nitrification, the aerobic conversion of ammonia into nitrates, is one of the most important functions in an aquaponics system as it reduces the toxicity of the water for fish, and allows the resulting nitrate compounds to be removed by the plants for nourishment.

Ammonia is steadily released into the water through the excreta and gills of fish as a product of their metabolism, but must be filtered out of the water since higher concentrations of ammonia (commonly between 0.5 and 1 ppm) can kill fish. Although plants can absorb ammonia from the water to some degree, nitrates are assimilated more easily, thereby efficiently reducing the toxicity of the water for fish. Ammonia can be converted into other nitrogenous compounds through healthy populations of:

- *Nitrosomonas*: bacteria that convert ammonia into nitrites, and
- *Nitrobacter*: bacteria that convert nitrites into nitrates.

In an aquaponics system, the bacteria responsible for this process form a biofilm on all solid surfaces throughout the system that are in constant contact with the water. The submerged roots of the vegetables combined have a large surface area, so that many bacteria can accumulate there. Together with the saliency of ammonia and nitrites in the water, the surface area determines the speed with which nitrification takes place. Care for these bacterial colonies is important as to regulate the full assimilation of ammonia and nitrite. This is why most aquaponics systems include a biofiltering unit, which helps facilitate growth of these microorganisms. Typically, after a system has stabilised ammonia levels range from 0.25 to 2.0 ppm; nitrite levels range from 0.25 to 1 ppm, and nitrate levels range from 2 to 150 ppm. During system startup, spikes may occur in the levels of ammonia (up to 6.0 ppm) and nitrite (up to 15 ppm), with nitrate levels peaking later in the startup phase. Since the nitrification process acidifies the water, non-sodium bases such as potassium hydroxide or calcium hydroxide can be added for neutralising the water's pH if insufficient quantities are naturally present in the water to provide a buffer against acidification. In addition, selected minerals or nutrients such as iron can be added in addition to the fish waste that serves as the main source of nutrients to plants.

A good way to deal with solids buildup in aquaponics is the use of worms, which liquefy the solid organic matter so that it can be utilised by the plants and/ or animals.

Hydroponics Subsystem

Plants are grown as in hydroponics systems, with their roots immersed in the nutrient-rich effluent water. This enables them to filter out the ammonia that is toxic to the aquatic animals, or its metabolites. After the water has passed through the hydroponic subsystem, it is cleaned and oxygenated, and can return to the aquaculture vessels. This cycle is continuous. Common aquaponic applications of hydroponic systems include:

- *Deep-water raft aquaponics*: styrofoam rafts floating in a relatively deep aquaculture basin in troughs.
- *Recirculating aquaponics*: solid media such as gravel or clay beads, held in a container that is flooded with water from the aquaculture. This type of aquaponics is also known as *closed-loop aquaponics*.
- *Reciprocating aquaponics*: solid media in a container that is alternately flooded and drained utilising different types of siphon drains. This type of aquaponics is also known as *flood-and-drain aquaponics* or *ebb-and-flow aquaponics*.
- Other systems use towers that are trickle-fed from the top, nutrient film technique channels, horizontal PVC pipes with holes for the pots, plastic

barrels cut in half with gravel or rafts in them. Each approach has its own benefits.

Most green leaf vegetables grow well in the hydroponic subsystem, although most profitable are varieties of chinese cabbage, lettuce, basil, roses, tomatoes, okra, cantaloupe and bell peppers. Other species of vegetables that grow well in an aquaponic system include beans, peas, kohlrabi, watercress, taro, radishes, strawberries, melons, onions, turnips, parsnips, sweet potato and herbs. Since plants at different growth stages require different amounts of minerals and nutrients, plant harvesting is staggered with seedings growing at the same time as mature plants. This ensures stable nutrient content in the water because of continuous symbiotic cleansing of toxins from the water.

Aquaculture Subsystem

Freshwater fish are the most common aquatic animal raised using aquaponics, although freshwater crayfish and prawns may also be used. In practice, tilapia are the most popular fish for home and commercial projects that are intended to raise edible fish, although barramundi, Silver Perch, Eel-tailed catfish or tandanus catfish, Jade perch and Murray cod are also used. For temperate climates when there isn't ability or desire to maintain water temperature, bluegill and catfish are suitable fish species for home systems. Koi and goldfish may also be used, if the fish in the system need not be edible.

Normal Operations

Aquaponic systems do not typically discharge or exchange water under normal operation, but instead recirculate and reuse water very effectively. The system relies on the relationship between the animals and the plants to maintain a stable aquatic environment that experience a minimum of fluctuation in ambient nutrient and oxygen levels. Water is only added to replace water loss from absorption and transpiration by plants, evapouration into the air from surface water, overflow from the system from rainfall, and removal of biomass such as settled solid wastes from the system. As a result, aquaponics uses approximately 2% of the water that a conventionally irrigated farm requires for the same vegetable production. This allows for aquaponic production of both crops and fish in areas where water or fertile land is scarce. Aquaponic systems can also be used to replicate controlled wetland conditions that are useful for water treatment by reclaiming potable water from typical household sewage. The nutrient-filled overflow water can be accumulated in catchment tanks, and reused to accelerate growth of crops planted in soil, or it may be pumped back into the aquaponic system to top up the water level..

The three main inputs to the system are water, feed given to the aquatic animals, and electricity to pump water between the aquaculture subsystem and the hydroponics subsystem. Spawn or fry may be added to replace grown fish that are taken out from the system to retain a stable system. In terms of outputs, an aquaponics system may continually yield plants such as vegetables grown in hydroponics, and edible aquatic species raised in an aquaculture.

History

Ancient

Aquaponics has ancient roots, although there is some debate on its first occurrence:

- Aztec cultivated agricultural islands known as *chinampas* and are considered by some as the first form of aquaponics for agricultural use where plants were raised on stationary (and sometime movable) islands in lake shallows and waste materials dredged from the Chinampa canals and surrounding cities are used to manually irrigate the plants.
- South China and Thailand who cultivated and farmed rice in paddy fields in combination with fish are cited as examples of early aquaponics. These polycultural farming systems existed in many Far Eastern countries and raised fish such as the oriental loach swamp eel, Common and crucian carp as well as pond snails in the paddies.

Regions

North America

United States: While the development of aquaponics is often attributed to the various works of the New Alchemy Institute and the works of Dr. Mark McMurtry et al. at the North Carolina State University, many papers of initial development of aquaponics concepts pre-date both institutions by nearly a decade.

Tom and Paula Speraneo, owners of a small greenhouse operation near West Plains, Missouri, modified the North Carolina State method and raised tilapia in above-ground tanks inside a solar greenhouse. The effluent from the tanks was used to fertigate gravel-cultured vegetables in raised benches.

In addition, the Speraneos manipulated the watering cycle, and this methodology of the Speranos forms the basis for the style of "flood & drain" media grow bed aquaponics systems that have been widely adopted in Australia based on models promoted by Joel Malcolm and Murray Hullam, and are now gaining increasing popularity in the United States.

Inspired by the successes of the New Alchemy Institute and the North Carolina State University with aquaponics, other institutes followed suit. Besides the reciprocating aquaponics based on the techniques developed by Dr. Mark McMurtry et al. at the North Carolina State University, Dr. James Rakocy and his colleagues at the University of the Virgin Islands researched and developed the "Deep Water" or "Raft Culture" aquaponics The system combines tilapia with various vegetables.

In 1997 Rebecca L. Nelson and John S. Pade began publishing the Aquaponics Journal, a quarterly scientific journal that brings together research and various applications of aquaponics from around the globe. In 2008, they wrote and published the first book on aquaponics, *Aquaponic Food Production.* Since then there have been numerous books, and "e-books" appear concerning aquaponics.

In 2010 the Aquaponic Gardening Community was started, which has now become the largest on-line gathering place for aquaponics enthusiasts in North America. Members of that community launched the Aquaponics Association in 2011 after the first Aquaponics Conference was held in Orlando, Florida in September of that year.

Recent years have seen a shift towards community integration of aquaponics, such as the nonprofit foundation Growing Power that offers Milwaukee youth job opportunities and training while growing food for their community. The model has spawned several satellite projects in other cities, such as New Orleans where the Vietnamese fisherman community has suffered from the Horizon Disaster. In addition, aquaponic gardeners from all around the world have gathered in online community sites and forums to openly share their experiences and move the development of this fantastic form of gardening forward.

Canada

The first aquaponics research in Canada was a small system added onto existing aquaculture research at a research station in Lethbridge. Canada saw a rise in aquaponics setups throughout the 90s, predominantly as commercial installations, that for example combine trout with floating lettuce production, or to water fruiting vegetable crops that warm up the water too much to be recirculated back into the fish ponds.

Eels are also known to be raised. A set-up based on the deep water system developed at the University of Virgin Islands was built in a greenhouse at Brooks, Alberta where Dr. Nick Savidov and colleagues researched aquaponics from a background of plant science.

The team made findings on rapid root growth in aquaponics systems, on closing the solid waste loop, and that because of certain advantages in the system over traditional aquaculture, the system can run well at a low pH level, which is favoured by plants but not fish.

The Edmonton Aquaponics Society in Northern Alberta is adapting Dr. Savidov's commercially sized system to a smaller scale prototype that can be operated by families, small groups, or restaurants. They intend to further develop the closed solid waste loop.

South America

Barbados

Barbados is a densely populated island that deals with water scarcity. In 40 years' time, focus has shifted from domestic fruit and vegetable production on small farms to importing 80% of all fruits and vegetables for cost reasons. Aquaponics would make Barbados and other Caribbean islands less dependent on the world food market and reduce stress on the dwindling fish supplies. An inter-organisational project that started in late 2009 sets out to encourage and enable Barbadians to start aquaponics at home, with revenue generated by selling produce to tourists.

Asia

Bangladesh

Bangladesh is the world's most densely populated country. Most of its farmers use/misuse agrochemicals to enhance food production/storage life. However, the country is yet to establish a good monitoring system to ensure chemical-residue free safe food in the market, capable city dwellers are looking for a better way to grow their own safe foods for domestic consumption. To help support mainly this client group and people living in adverse climatic conditions such as salinity-prone southern part and flood-pronehaor area in the eastern region, a low-cost model of backyard aqaponics system has been developed by Professor Dr. M.A. Salam at the Department of Aquaculture of Bangladesh Agricultural University, Mymensingh. The initial model has been developed in 2011, and demonstrated and reported to the press in January 2012. Further researches may be necessary to develop appropriate models for different situations among client groups.

Taiwan

Taiwan is a densely populated island that is faced with freshwater scarcity. Dispensation of water is government-controlled. The closed-loop system of aquaponics is used by agricultural farmers to save water by also rearing fish, while fish farmers grow plants that filter the water from the fish tanks. Taiwan Aquaponics Association is the not-for profit federation to promote Aquaponics at Taiwan.

Australia

Tom and Paula Speraneo, owners of a small greenhouse operation near West Plains, Missouri, modified the North Carolina State method developed by Dr. Mark McMurtry et al., and raised tilapia in above-ground tanks inside a solar greenhouse. The effluent from the tanks was used to fertigate gravel-cultured vegetables in raised benches. In addition, the Speraneos manipulated the watering cycle, and this methodology of the Speranos forms the basis for the style of "flood

& drain" media grow bed aquaponics systems that have been widely adopted in Australia based on models promoted by Joel Malcolm and Murray Hullam, and are now gaining increasing popularity in the United States.

Due to a ban on tilapia in all states except for Western Australia, native freshwater fish including Silver Perch, Jade Perch, Sleepy Cod, Murray Cod and Barramundi are popular in aquaponics and aquaculture systems, along with non-native Rainbow Trout, Brown Trout and Crayfish such as common Yabby and Redclaw.

Pros and Cons

The unique advantages of aquaponic systems are:

- Conservation through constant water reuse and recycling.
- Organic fertilization of plants with natural fish emulsion.
- The elimination of solid waste disposal from intensive aquaculture.
- The reduction of needed cropland to produce crops.
- The overall reduction of the environmental footprint of crop production.
- Building small efficient commercial installations near markets reduces food miles.
- Reduction of pathogens that often plague aquaculture production systems.

Some conceivable disadvantages with aquaponics are:

- Initial expenses for housing, tank, plumbing, pumps, and grow beds.
- The infinite number of ways in which a system can be configured lends itself to equally varying results, conflicting research, and successes or failures.
- Some aquaponic installations rely heavily on man-made energy, technological solutions, and environmental control to achieve recirculation and water/ambient temperatures. However, if a system is designed with energy conservation in mind, using alternative energy and a reduced number of pumps by letting the water flow downwards as much as possible, it can be highly energy efficient.
- While careful design can minimise the risk, aquaponics systems can have multiple 'single points of failure' where problems such as an electrical failure or a pipe blockage can lead to a complete loss of fish stock.
- Like all aquaculture based systems, stock feed usually consists of fish meal derived from lower value species. Ongoing depletion of wild fish stocks makes this practice unsustainable. Organic fish feeds may prove to be a viable alternative that negates this concern. Other alternatives include growing duckweed with an aquaponics system that feeds the same fish

grown on the system, excess worms grown from vermiculture composting, as well as growing black soldier fly larvae to feed to the fish using composting grub growers.

Copper Alloys in Aquaculture

Recently, copper alloys have become important netting materials in aquaculture (the farming of aquatic organisms including fish farming). Various other materials including nylon, polyester, polypropylene, polyethylene, plastic-coated welded wire, rubber, patented twine products (Spectra, Dyneema), and galvanised steelare also used for netting in aquaculture fish enclosures around the world. All of these materials are selected for a variety of reasons, including design feasibility, material strength, cost, and corrosion resistance.

What sets copper alloys apart from the other materials used in fish farming is that copper alloys are antimicrobial, that is, they destroy bacteria, viruses, fungi, algae, and other microbes.

In the marine environment, the antimicrobial/algaecidal properties of copper alloys prevent biofouling, which can briefly be described as the undesirable accumulation, adhesion, and growth of microorganisms, plants, algae, tubeworms, barnacles, mollusks, and other organisms on man-made marine structures. By inhibiting microbial growth, copper alloy aquaculture pens avoid the need for costly net changes that are necessary with other materials. The resistance of organism growth on copper alloy nets also provides a cleaner and healthier environment for farmed fish to grow and thrive.

In addition to their antifouling benefits, copper alloys have strong structural and corrosion-resistant properties in marine environments.

It is the combination of all of these properties – antifouling, high strength, and corrosion resistance – that has made copper alloys a desirable material for such marine applications as condenser tubing, water intake screens, ship hulls, offshore structure, and sheathing. In the past 25 years or so, the benefits of copper alloys have caught the attention of the marine aquaculture industry. The industry is now actively deploying copper alloy netting and structural materials in commercial large-scale fish farming operations around the world.

Importance of Aquaculture

Much has been written about the degradation and depletion of natural fish stocks in rivers, estuaries, and the oceans (Overfishing). Because industrial fishing has become extremely efficient, ocean stocks of large fish, such as tuna, cod, and halibut have declined by 90% in the past 50 years.

Aquaculture, an industry that has emerged only in recent decades, has become one of the fastest growing sectors of the world food economy. Aquaculture already supplies more than half of the world's demand for fish. This percentage is predicted to increase dramatically over the next few decades.

The Problem of Biofouling

Biofouling is one of the most important problems in aquaculture. Biofouling occurs on non-copper materials in the marine environment, including fish pen surfaces and nettings. For example, it was noted that the open area of a mesh immersed for only seven days in a Tasmanian aquaculture operation decreased by 37% as a result of biofouling.

The biofouling process begins when algae spores, marine invertebrate larvae, and other organic material adhere to surfaces submerged in marine environments (e.g., fish nets in aquaculture). Bacteria then encourage the attachment of secondary unwanted colonisers.

Biofouling has strong negative impacts on aquaculture operations. Water flow and dissolved oxygen are inhibited due to clogged nets in fish pens. The end result is often diseased fish from infections, such as netpen liver disease, amoebic gill disease, and parasites. Other negative impacts include increased fish mortalities, decreased fish growth rates, premature fish harvesting, reduced fish product values and profitability, and an adversely impacted environment near fish farms.

Biofouling adds enormous weight to submerged fish netting. Two hundredfold increases in weight have been reported. This translates, for example, to two thousand pounds of unwanted organisms adhered to what was once a clean 10-pound fish pen net. In South Australia, biofouling weighing 6.5 tonnes (approximately 13,000 pounds) was observed on a fish pen net. This extra burden often results in net breakage and additional maintenance costs.

To combat parasites from biofouling in finfish aquaculture, treatment protocols such as cypermethrin, azamethiphos, and emamectin benzoate may be administered, but these have been found to have detrimental environmental effects, for example, in lobster operations.

To treat diseases in fish raised in biofouled nets, fish stocks are administered antibiotics. The antibiotics can have unwanted long-term health effects on consumers and on coastal environments near aquaculture operations. To combat biofouling, operators often implement costly maintenance measures, such as frequent net changing, cleaning/removal of unwanted organisms from nets, net repairs, and chemical treatment including antimicrobial coatings on nylon nets. The cost of antifouling a single salmon net can be several thousand British pounds. In some sectors of the European aquaculture industry, cleaning biofouled fish and shellfish pens can cost 5–20% of its market value. Heavy fouling can reduce the saleable product in nets by 60–90%. Antifouling coatings are often used on nylon nets because the process is more economical than manual cleaning. When nylon nets are coated with antifouling compounds, the coatings repel biofouling for a period of time, usually between several weeks to several months. However, the nets eventually succumb to biofouling. Antifouling coatings containing cuprous oxide algaecide/biocide are the coatings technology used almost exclusively in the fish farming industry today. The treatments usually flake off within a few weeks to six to eight months.

Biofouled nets are replaced after several months of service, depending on environmental conditions, in a complicated, costly, and labour-intensive operation that involves divers and specialised personnel. During this process, live fish in nets must be transferred to clean pens, which causes undue stress and asphyxiation that results in some loss of fish. Biofouled nets that can be reused are washed on land via manual brushing and scrubbing or high-pressure water hosing. They are then dried and re-impregnated with antifouling coatings.

A line of net cleaners is available for in-situ washings where permitted. But, even where not permitted by environmental, fisheries, maritime, and sanitary authorities, should the lack of dissolved oxygen in submerged pens create an emergency condition that endangers the health of fish, divers may be deployed with special in situ cleaning machinery to scrub biofouled nets.

The aquaculture industry is addressing the negative environmental impacts from its operations. As the industry evolves, a cleaner, more sustainable aquaculture industry is expected to emerge, one that may increasingly rely on materials with anti-fouling, anti-corrosive, and strong structural properties, such as copper alloys.

Antifouling Properties of Copper Alloys

In the aquaculture industry, sound animal husbandry translates to keeping fish clean, well fed, healthy, and not overcrowded. One solution to keeping farmed fish healthy is to contain them in antifouling copper alloy nets and structures.

Researchers have attributed copper's resistance to biofouling, even in temperate waters, to two possible mechanisms: 1) a retarding sequence of colonisation through release of antimicrobial copper ions, thereby preventing the attachment of microbial layers to marine surfaces; and, 2) separating layers that contain corrosive products and the spores of juveniles or macro-encrusting organisms.

The most important requirement for optimum biofouling resistance is that the copper alloys should be freely exposed or electrically insulated from less noble alloys and fromcathodic protection. Galvanic coupling to less noble alloys and cathodic protection prevent copper ion releases from surface films and therefore reduce biofouling resistance.

As temperatures increase and water velocities decrease in marine waters, biofouling rates dramatically rise. However, copper's resistance to biofouling is observed even in temperate waters. Studies in La Herradura Bay, Coquimbo,

Chile, where biofouling conditions are extreme, demonstrated that a copper alloy (90% copper, 10% nickel) avoided macro-encrusting organisms.

Corrosion Behaviour of Copper Alloys

Copper alloys used in sea water service have low general corrosion rates but also have a high resistance to many localised forms of corrosion. A technical discussion regarding various types of corrosion, application considerations (e.g., depth of installations, effect of polluted waters, sea conditions), and the corrosion characteristics of several copper alloys used in aquaculture netting is available (i.e., copper-nickel, copper-zinc, and copper-silicon).

Environmental Performance of Copper Alloy Mesh

Many complicated factors influence the environmental performance of copper alloys in aquaculture operations. A technical description of antibiofouling mechanisms, fish health and welfare, fish losses due to escapes and predator attacks, and reduced life cycle environmental impacts is summarised in this reference.

Types of Copper Alloys

Copper-zinc brass alloys are currently (2011) being deployed in commercial-scale aquaculture operations in Asia, South America and the USA (Hawaii). Extensive research, including demonstrations and trials, are currently being implemented on two other copper alloys: copper-nickel and copper-silicon. Each of these alloy types has an inherent ability to reduce biofouling, pen waste, disease, and the need for antibiotics while simultaneously maintaining water circulation and oxygen requirements. Other types of copper alloys are also being considered for research and development in aquaculture operations.

The University of New Hampshire is in the midst of conducting experiments under the auspices of the International Copper Association (ICA) to evaluate the structural, hydrodynamic, and antifouling response of copper alloy nets. Factors to be determined from these experiments, such as drag, pen dynamic loads, material loss, and biological growth – well documented for nylon netting but not fully understood for copper-nickel alloy nets – will help to design fish pen enclosures made from these alloys. The East China Sea Fisheries Research Institute, in Shanghai, China, is also conducting experimental investigations on copper alloys for ICA.

Copper-zinc Alloys

The Mitsubishi-Shindoh Co., Ltd., has developed a proprietary copper-zinc brass alloy, called UR30, specifically designed for aquaculture operations. The alloy, which is composed of 64% copper, 35.1% zinc, 0.6% tin, and 0.3% nickel, resists mechanical abrasion when formed into wires and fabricated into chain link, woven, or other types of flexible mesh. Corrosion rates depend on the depth of submersion and seawater conditions. The average reported corrosion rate reported for the alloy is <5 ìm/yr based on 2- and 5-year exposure trials in seawater.

The Ashimori Industry Company, Ltd., has installed approximately 300 flexible pens with woven chain link UR30 meshes in Japan to raise Seriola (i.e., yellowtail, amberjack, kingfish, hamachi). The company has installed another 32 brass pens to raise Atlantic salmon at the Van Diemen Aquaculture operations in Tasmania, Australia. In Chile, EcoSea Farming S.A. has installed a total of 62 woven chain link brass mesh pens to raise troutand Atlantic salmon. In Panama, China, Korea, Turkey, and the USA, demonstrations and trials are underway using flexible pens with woven chain link UR30 and other mesh forms and a range of copper alloys.

To date, in over 10 years of aquaculture experience, chain link mesh fabricated by these brass alloys have not suffered from dezincification, stress corrosion cracking, or erosion corrosion.

Copper-nickel Alloys

Copper-nickel alloys were developed specifically for seawater applications over five decades ago. Today, these alloys are being investigated for their potential use in aquaculture.

Copper-nickel alloys for marine applications are usually 90% copper, 10% nickel, and small amounts of manganese and iron to enhance corrosion resistance. The seawater corrosion resistance of copper-nickel alloys results in a thin, adherent, protective surface film which forms naturally and quickly on the metal upon exposure to clean seawater.

The rate of corrosion protective formation is temperature dependent. For example, at 27 °C (i.e., a common inlet temperature in the Middle East), rapid film formation and good corrosion protection can be expected within a few hours. At 16 °C, it could take 2–3 months for the protection to mature. But once a good surface film forms, corrosion rates decrease, normally to 0.02–0.002 mm/yr, as protective layers develop over a period of years. These alloys have good resistance to chloride pitting and crevice corrosion and are not susceptible to chloride stress corrosion.

Copper-silicon Alloys

Copper-silicon has a long history of use as screws, nuts, bolts, washers, pins, lag lag bolts, and staples in wooden sailing vessels in marine environments. The alloys are often composed of copper, silicon, and manganese. The inclusion of silicon strengthens the metal.

As with the copper-nickel alloys, corrosion resistance of copper-silicon is due to protective films that form on the surface over a period of time. General corrosion rates of 0.025–0.050mm have been observed in quiet waters. This rate decreases towards the lower end of the range over long-term exposures (e.g., 400–600 days). There is generally no pitting with the silicon-bronzes.

Also there is good resistance to erosion corrosion up to moderate flow rates. Because copper-silicon is weldable, rigid pens can be constructed with this material. Also, because welded copper-silicon mesh is lighter than copper-zinc chain link, aquaculture enclosures made with copper-silicon may be lighter in weight and therefore a potentially less expensive alternative.

Luvata Appleton, LLC, is researching and developing a line of copper alloy woven and welded meshes, including a patent-pending copper silicon alloy, that are marketed under the trade name Seawire. Copper-silicon alloy meshes have been developed by the firm to raise various marine organisms in test trials that are now in various stages of evaluation. These include raising cobia in Panama, lobsters in Maine, and crabs in the Chesapeake Bay. The company is working with various universities to study its material, including the University of Arizona to study shrimp, the University of New Hampshire to study cod, and Oregon State Universityto study oysters. Results will be provided when the studies are concluded.

Fisheries Science

Figure: Fisheries scientists sorting a catch of small fish and Norway lobster

Fisheries science is the academic discipline of managing and understanding fisheries. It is a multidisciplinary science, which draws on the disciplines of limnology, oceanography, freshwater biology, marine biology, conservation, ecology, population dynamics, economics and management to attempt to provide an integrated picture of fisheries. In some cases new disciplines have emerged, as in the case of bioeconomics.

Fisheries science is typically taught in a university setting, and can be the focus of an undergraduate, master's or Ph.D. program. Some universities offer fully integrated programs in fisheries science.

Notable Contributors

- Spencer F. Baird – founding scientist of the United States Fish Commission
- Fedor I. Baranov - Russian scientist and the father of the *Baranov catch equation*, thereby being one of the founding fathers of fisheries science
- Ludwig von Bertalanffy – Austrian-born biologist and a founder of general systems theory
- Ray Beverton – English fisheries biologist; known for the Beverton–Holt model (with Sidney Holt), cred with being one of the founders of fisheries science
- Villy Christensen – fisheries scientist and ecosystem modeler, known for his work on the development of Ecopath
- John N. Cobb – founder of the first college of fisheries in the United States, the University of Washington College of Fisheries, in 1919
- David Cushing – English born fisheries biologist, who is cred with the development of the match/mismatch hypothesis
- W. Harry Everhart – American fisheries scientist, educator, administrator, and author of several widely used fisheries texts
- Rainer Froese – Known for his work on the development and coordination of FishBase
- Gotthilf Hempel – German marine biologist and oceanographer, and co-founder of the Alfred Wegener Institute for Polar and Marine Research
- Walther Herwig – Prussian lawyer and promoter of high seas fishing and research
- Ray Hilborn – Canadian-born fisheries biologist, with strong contributions towards fisheries management
- Johan Hjort – Norwegian fisheries biologist, marine zoologist, and oceanographer
- Bruno Hofer – German fishery scientist, cred with being the founder of fish pathology
- Sidney Holt – English fisheries biologist; known for the Beverton–Holt model (with Ray Beverton), cred with being one of the founders of fisheries science
- Uwe Kils – German marine biologist specialising in planktology. Inventor of the ecoSCOPE

- Robert T. Lackey – Canadian born fisheries scientist and political scientist known for his work involving the role of science in policy making
- Leo Margolis – Canadian parasitologist and head of the Pacific Biological Station in Nanaimo, British Columbia
- R. J. McKay – Australian-born biologist and a specialist in translocated freshwater fishes
- Ransom A. Myers – Canadian marine biologist and conservationist
- Daniel Pauly – prominent French-born fisheries scientist, known for his work studying human impacts on global fisheries
- Tony J. Pitcher – known for work on the impacts of fishing, management appraisals and the shoaling behaviour of fish
- Michael A. Rice – American known for work on molluscan fisheries
- Bill Ricker – Canadian fisheries biologist, known for the Ricker model, cred with being one of the founders of fisheries science
- Ed Ricketts – a colourful American marine biologist and philosopher who introduced ecology to fisheries science.
- Callum Roberts – British marine conservation biologist, known for his work on the role marine reserves play in protecting marine ecosystems
- Harald Rosenthal – German hydrobiologist known for his work in fish farming and ecology
- Carl Safina – author of several writings on marine ecology and the ocean
- Georg Ossian Sars – Norwegian marine biologist cred with the discovery of a number of new species and known for his analysis of cod fisheries
- Tore Schweder – Norwegian statistician whose work includes the assessment of marine resources
- Milner Baily Schaefer – notable for his work on the population dynamics of fisheries
- Ussif Rashid Sumaila – noted for his analysis of the economic aspects of fisheries
- Fred Utter – noted as the founding father of the field of fishery genetics
- Carl Walters – American born biologist known for his work involving fisheries stock assessments, the adaptive management concept, and ecosystem modelling

Fish Hatchery

A fish hatchery is a "place for artificial breeding, hatching and rearing through the early life stages of animals, finfish and shellfish in particular".

Hatcheries produce larval and juvenile fish (and shellfish and crustaceans) primarily to support the aquaculture industry where they are transferred to on-growing systems i.e. fish farms to reach harvest size. Some species that are commonly raised in hatcheries include Pacific oysters, shrimp, Indian prawns, salmon, tilapia andscallops. The value of global aquaculture production is estimated to be US$98.4 billion in 2008 with China significantly dominating the market, however the value of aquaculture hatchery and nursery production has yet to be estimated. Additional hatchery production for small-scale domestic uses, which is particularly prevalent in South-East Asia or for conservation programmes, has also yet to be quantified.

There is much interest in supplementing exploited stocks of fish by releasing juveniles that may be wild caught and reared in nurseries before transplanting, or produced solely within a hatchery. Culture of finfish larvae has been utilised extensively in the United States in stock enhancement efforts to replenish natural populations. The U.S. Fish and Wildlife Service have established a National Fish Hatchery System to support the conservation of native fish species.

Purpose

Hatcheries produce larval and juvenile fish and shellfish for transferal to aquaculture facilities where they are 'on-grown' to reach harvest size. Hatchery production confers three main benefits to the industry;

1. *Out of season production:* Consistent supply of fish from aquaculture facilities is an important market requirement. Broodstock conditioning can extend the natural spawning season and thus the supply of juveniles to farms. Supply can be further guaranteed by sourcing from hatcheries in the opposite hemisphere i.e. with opposite seasons.
2. *Genetic improvement:* Genetic modification is conducted in some hatcheries to improve the quality and yield of farmed species. Artificial fertilisation facilitates selective breeding programs which aim to improve production characteristics such as growth rate, disease resistance, survival, colour, increased fecundity and/or lower age of maturation. Genetic improvement can be mediated by selective breeding, via hybridisation, or other genetic manipulation techniques.
3. *Reduce dependence on wild-caught juveniles:* In 2008 aquaculture accounted for 46% of total food fish supply, around 115 million tonnes. Although wild caught juveniles are still utilised in the industry, concerns over sustainability of extracting juveniles, and the variable timing and magnitude of natural spawning events, make hatchery production an attractive alternative to support the growing demands of aquaculture.

3

Production Steps

Broodstock

Broodstock conditioning is the process of bringing adults into spawning condition by promoting the development of gonads. Broodstock conditioning can also extend spawning beyond natural spawning periods, or for production of species reared outside their natural geographic range with different environmental conditions.

Some hatcheries collect wild adults and then bring them in for conditioning whilst others maintain a permanent breeding stock. Conditioning is achieved by holding broodstock in flow-through tanks at optimal conditions for light, temperature, salinity, flow rate and food availability (optimal levels are species specific). Another important aspect of broodstock conditioning is ensuring the production of high quality eggs to improve growth and survival of larvae by optimising the health and welfare of broodstock individuals. Egg quality is often determined by the nutritional condition of the mother. High levels of lipid reserves in particular are required to improve larval survival rates.

Spawning

Natural spawning can occur in hatcheries during the regular spawning season however where more control over spawning time is required spawning of mature animals can be induced by a variety of methods. Some of the more common methods are:

Manual Stripping *:* For shellfish, gonads are generally removed and gametes are extracted or washed free. Fish can be manually stripped of eggs and sperm by stroking the anaesthetised fish under the pectoral fins towards the anus causing gametes to freely flow out.

Environmental Manipulation: Thermal shock, where cool water is alternated with warmer water in flow-through tanks can induce spawning. Alternatively, if environmental cues that stimulate natural spawning are known, these can be mimicked in the tank e.g. changing salinity to simulate migratory behaviour. Many individuals can be induced to spawn this way, however this increases the likelihood of uncontrolled fertilisation occurring.

Chemical Injection: A number of chemicals can be used to induce spawning with various hormones being the most commonly used.

Fertilisation

Prior to fertilisation, eggs can be gently washed to remove wastes and bacteria that may contaminate cultures. Promoting cross-fertilisation between a large number of individuals is necessary to retain genetic diversity in hatchery produced stock. Batches of eggs are kept separate, fertilised with sperm obtained from several males and allowed to stand for an hour or two before samples are analysed under a microscope to ensure high rates of fertilisation and to estimate numbers to be transferred to larval rearing tanks.

Larvae

Rearing larvae through the early life stages is conducted in nurseries which are generally closely associated with hatcheries for fish culture whilst it is common for shellfish nurseries to exist separately.

Nursery culture of larvae to rear juveniles of a size suitable for transferral to on-growing facilities can be performed in a variety of different systems which may be entirely land-based, or larvae may be later transferred to sea-based rearing systems which reduce the need to supply feed. Juvenile survival is dependant on very high quality water conditions. Feeding is an important component of the rearing process.

Although many species are able to grow on maternal reserves alone (lecithotrophy), most commercially produced species require feeding to optimise survival, growth, yield and juvenile quality. Nutritional requirements are species specific and also vary with larval stage. Carnivorous fish are commonly fed with live prey; rotifers are usually offered to early larvae due to their small size, progressing to larger *Artemia* nauplii or zooplankton. The production of live feed on-site or buying-in is one of the biggest costs for hatchery facilities as it is a labour intensive process. The development of artificial feeds is targeted to reduce the costs involved in live feed production and increase the consistency of nutrition, however decreased growth and survival has been found with these alternatives.

Settlement of Shellfish

The hatchery production of shellfish also involves a crucial settling phase where free-swimming larvae settle out of the water onto a substrate and undergo metamorphosis if suitable conditions are found. Once metamorphosis has taken place the juveniles are generally known as spat, it is this phase which is then transported to on-growing facilities. Settlement behaviour is governed by a range of cues including substrate type, water flow, temperature, and the presence of chemical cues indicating the presence of adults, or a food source etc. Hatchery facilities therefore need to understand these cues to induce settlement and also be able to substitute artificial substrates to allow for easy handling and transportation with minimal mortality.

Hatchery Design

Hatchery designs are highly flexible and are tailored to the requirements of site, species produced, geographic location, funding and personal preferences. Many hatchery facilities are small and coupled to larger on-growing operations, whilst others may produce juveniles solely for sale. Very small-scale hatcheries are often utilised in subsistence farming to supply families or communities particularly in south-east Asia. A small-scale hatchery unit consists of larval rearing tanks, filters, live food production tanks and a flow through water supply. A generalised commercial scale hatchery would contain a broodstock holding and spawning area, feed culture facility, larval culture area, juvenile culture area, pump facilities, laboratory, quarantine area, and offices and bathrooms.

Expense

Labour is generally the largest cost in hatchery production making up more that 50% of total costs. Hatcheries are a business and thus economic viability and scale of production are vital considerations. The cost of production for stock-enhancement programmes is further complicated by the difficulty of assessing the benefits to wild populations from restocking activities.

Issues

Genetic

Hatchery facilities present three main problems in the field of genetics. The first is that maintenance of a small number of broodstock can cause inbreeding and potentially lead to inbreeding depression thus affecting the success of the facility. Secondly, hatchery reared juveniles, even from a fairly large broodstock, can have greatly reduced genetic diversity compared to wild populations (the situation is comparable to the founder effect). Such fish that escape from farms or are released for restocking purposes may adversely affect wild population genetics and viability. This is of particular concern where escaped fish have been actively bred or are otherwise genetically modified. The third key issue is that genetic modification of food items is highly undesirable for many people. Genetically modified food controversies

Fish Farms

Other arguments that surround fish farms such as the supplementation of feed from wild caught species, the prevalence of disease, fish welfare issues and potential effects on the environment are also issues for hatchery facilities.

Mount Whitney Fish Hatchery

The Mount Whitney Fish Hatchery, located in Independence, California, in the United States, is an historic fish hatchery that has played an important role in the preservation of the golden trout, California's state fish.

Construction

The facility was built and operated by the California State Fish & Game Commission, now known as the California Department of Fish and Game. Starting in 1915, the citizens of Independence began a local fundraising drive to purchase a site for a proposed state fish hatchery. $1,500.00 was raised, and an ideal 40-acre (16 ha) site was purchased on Oak Creek, just north of the town. Fish and Game Commissioner M. J. Connell instructed the design team led by Charles Dean of the State Department of Engineering "to design a building that would match the mountains, would last forever, and would be a showplace for all time." The architectural style they chose is Tudor Revival. Construction began in March, 1916, with a final budget of approximately $60,000.00. The walls of the building are constructed using 3,500 short tons (3,200 t) of native granite collected within a quarter mile (400 m) of the site. The walls are two to three feet (600 to 900 mm) thick. The roof is red Spanish tile made in Lincoln, California.

Operation

When construction was completed in 1917, it was the largest and best equipped hatchery in California and could produce 2,000,000 fish fry per year. Initially, fish eggs were collected from the Rae Lakes and were transported to the hatchery by mule train. Starting in 1918, golden trout eggs were collected from the Cottonwood Lakes, and production of golden trout continued until 2008. This program was the sole source of California golden trout eggs for 90 years.

In July 1931, the Mount Whitney Fish Hatchery and the Colorado Fish Commission traded 30,000 Colorado River cutthroat trout eggs for 25,000 golden trout eggs. The resulting Colorado cutthroat fry were planted in remote High Sierra lakes at very high elevations. Over the following 50 years, the population in Colorado became endangered due to habitat destruction and interbreeding with other species of trout. The cutthroats now living in California remained pure. In

1987, California and Colorado cooperated to transplant 50 genetically pure cutthroats back to a remote lake in Rocky Mountain National Park, where they thrived.

Proposed Closure

In 1996, the California Department of Fish and Game proposed closing the hatchery due to budget cuts. Local and statewide opposition to the closure developed, and instead, a plan was approved to save the facility in order to "provide the public with an interpretation of the historical significance of the hatchery, knowledge of the hatchery's function and an understanding of our natural resources". Legislation was passed designating the "Friends of the Mt. Whitney Fish Hatchery" as a private group authorised to lease part of the hatchery in order to maintain and preserve it, in coordination with the State Office of Historic Preservation.

Wildfire and Mudslide

On July 5, 2007, a 55,000-acre (22,000 ha) wildfire burned upstream to the west of the Mount Whitney Fish Hatchery. As a result, a year later, on July 12, 2008, a heavy thunderstorm caused a massive mudslide in the fire-scarred Oak Creek watershed that swept downstream, severely damaging the ponds and water supplies of the hatchery, as well as two employee housing units. The main building escaped major damage. The Friends of the Mt. Whitney Fish Hatchery organised restoration work that allowed the interpretive centre and display pond to re-open on May 30, 2009. However, the future of full-scale hatchery production is uncertain.

The first group of fish to come out of the hatchery in three years were planted in Diaz Lake, and a Kids Fishing Day was held in May, 2010. An expanded interpretive centre is in the planning stages.

Raceway (Aquaculture)

A raceway, also known as a flow-through system, is an artificial channel used in aquaculture to culture aquatic organisms. Raceway systems are among the earliest methods used for inland aquaculture. A raceway usually consists of rectangular basins or canals constructed of concrete and equipped with an inlet and outlet. A continuous water flow-through is maintained to provide the required level of water quality, which allows animals to be cultured at higher densities within the raceway. Freshwater species such as trout, catfish and tilapia are commonly cultured in raceways. Raceways are also used for some marine species which need a constant water flow, such as juvenile salmon, brackish water seabass and seabream and marine invertebrates such as abalone.

Site Selection

The most important factor to consider when selecting a site for a raceway farm is the water supply. Water sources for raceway aquaculture operations are usually streams, springs, reservoirs or deep wells. Trout do best in spring water because it keeps a constant temperature, while catfish need a strong flow, about 80 litres per second for every 0.4 hectares of raceway. A backup water supply should be positioned so, if the water supply or pump fails, it can flow by gravity into the start of the raceway.

Construction

Most raceways are made of reinforced concrete, though some earthen raceways are also built. Earthen raceways with plastic liners cost little and are easy to build, but cleaning and disinfecting them is difficult and plastic linings are fragile. Reinforced concrete is more expensive, but is durable and can be shaped in complex ways. Raceway tanks can also be built from polyester resin. These tanks have smooth walls, and are mobile and easy to service. However, their cost limits them to small sizes, under 5 cubic metres.

Size

A raceway is most often a rectangular canal with a water current flowing from a supply end to an exit end. The length to width ratio is important in raceways. To prevent the fish stock from swimming in circular movements, which would cause debris to build up in the centre, a length to width ratio of at least six to one is recommended. If the width is too large this could result in a feeble current speed which is not desirable. The length of a raceway unit is usually constrained by the water quality or by how much stock a unit can hold for ease of management. The average depth of a raceway for fin fish, such as rainbow trout, is about one metre. This means each section in a raceway should be about 30 m long and 2.5–3 m wide. The landscape should sloped to one or two percent, so the flow through the system can be maintained by gravity. The raceway should not be curved, so the flow will be uniform.

A raceway farm for freshwater fin fish usually has a dozen or more parallel raceway strips build alongside each other, with each strip consisting of 15 to 20 or more serial sections. The risk of unhygienic conditions increases towards the lower level sections, and can be kept in check by ensuring there are not too many sections and the water flow is adequate. In order to isolate any diseased section and avoid transmitting the disease back to the upper raceways, each section should have its own drainage channel. Controls, such as weirs, are also needed to ensure individual raceways can't accidentally overflow or empty.

Water Flow

The water flow rate in a raceway system needs to be sufficiently high to meet the respiratory (dissolved oxygen) requirements for the species concerned and to flush out metabolic wastes, especially ammonia. In a well designed system, the existing water in the raceway is largely replaced by new water when the same volume of new water enters the raceway. Self cleaning can sometimes be achieved if the fish stocks density is sufficiently high and the water level is sufficiently low. For example, if trout are stocked at 20 kilograms per cubic metre, they can keep the raceway unit clean by their swimming movements, preventing waste solids from settling to the raceway floor.

However, in most cases it is necessary to frequently clean raceways. The simplest way is to lower the water level in the raceway units, which increases the speed of the water current, and then herd the fish together till they flush the waste from the raceway. Solid wastes which accumulate at the raceway bottom can be removed by pumps. Oxygen levels in the water can be keep high if the raceway units are placed one after the other with intermediate drops over weirs, or by the use of aeration systems such as pumps, blowers and agitators.

Generally the water should be replaced about every hour. This means a typical raceway section requires a flow rate around 30 litres per second. However, the optimum flow through rate depends on the species, because there are differences in the rates at which oxygen is consumed and metabolic wastes are produced. For example, trout and juvenile salmon are less tolerant of degraded water quality and require a more rapid water turnover than catfish or tilapia. The flow rate necessary to maintain water quality can also change through the year, as the temperature changes and the cultured species grow larger. For reason such as these, continuous monitoring of water quality is important, including measurements of water flow rates, pH levels and temperature, as well as the levels of dissolved oxygen, and suspended and solid waste material.

Maximum Load

The maximum load of organisms that can be cultured in a raceway system depends on the species, and particularly on the size of the species. For trout, stocking rates of 30 to 50 kg/m^3 are normal at the end of a rearing cycle, while for marine species, such as sea-bass and sea-bream, the achievable load is lower, between 15 and 20 kg/m^3. The total volume required for a raceway is calculated by dividing the total amount of fish in kg by the desired stocking rate in kg per m^3.

Feed

In most raceway aquaculture food needs to be supplied. The composition of the food, and the amount and time of feeding needs to be adjusted to the specific species. This can be optimised to reduce costs and minimise the amount of waste.

Waste Water

The treatment of waste water issuing from raceway farms is a major concern. Fish fecal matter and uneaten feed are typically the major elements of solid waste produced in raceway aquaculture farms. These can adversely impact the environment in the receiving water body. Of particular environmental concern is the waste product phosphorus. Excessive discharge of phosphorus to receiving waters can result ineutrophication. For example, in Korea poor waste treatments in trout farms resulted in reservoirs and rivers developing red tides, which caused wider social problems.

Because raceway aquaculture operations discharge large volumes of water, the concentration of discharged solids is low. This means it is not easy to treat and implement practical, cost effective treatments. Technologies for the removal of solids include microscreens, dual-drain tanks, swirl separators, plate separators, baffles, media filters, air flotation, foam fractionation, chemical flocculation, and constructed wetlands. But because of the impracticality and / or high costs of these methods, most of them are not applicable for commercial aquaculture. As a consequence, sedimentation (settling) is still the most widely applied and cost effective technology. From 1999, regulations in South Korean require that all raceway farms provide waste water treatment facilities covering at least 20% of the farmed area to prevent pollution of the freshwater environment.

Microponics

Microponics is the symbiotic integration of fish, plants and micro-livestock in a semi-controlled environment.

History

Microponics is a term created recently by Australian urban farmer, Gary Donaldson to describe his integrated backyard food production concept.

While Microponics was also the name given to an obscure grafting method used in hydroponics, Donaldson's use of the term was derived from the integration of micro-livestock (and micro-farming) and the production of fish and plants - aquaponics.

Microponics has its roots in the integrated aquaculture work undertaken by the New Alchemy Institute (Todd, McLarney et al.) during the late 1960s and early 1970s.

The New Alchemists had developed useful food production models based on the integration of fish, plants, ducks, rabbits and other organisms - all of which were housed in their solar and wind-powered Cape Cod Ark bio-shelter.

Integrated aquaculture, in which the byproducts (waste) of one species are converted to become the feedstock (food, fertilizer, etc.) of another species, struck an immediate chord with Gary Donaldson when he was first introduced to it in the mid-1970s.

His goal was to develop a backyard food production regime that would enable the average householder to grow their own clean fresh food through the application of modest skills and appropriate technology.

He reasoned that the one thing that all plant and animal species had in common was a need for water so it seemed logical that aquaculture should become a cornerstone of any integrated backyard food production system.

Once the New Alchemists had reconnected him with the notion of integration (which is largely traditional mixed farming with an appropriate technology twist), Donaldson still grappled with the issue of scale. At just under 30 metres long and 6 metres high, the Ark was rather too large for the average backyard.

Notwithstanding the innovative approach of the New Alchemists, translation of their work into an Australian context back in the 1970s was also hindered by the lack of information about the culture of Australian fish species.

In the ensuing twenty years, aquaculture in Australia came of age and local researchers identified a number of suitable freshwater aquaculture species.

Since 2007, Gary Donaldson has satisfied himself that limited quantities (in a backyard context) of Australian freshwater fish could be successfully grown out in as little as 600 litres of water.

Integrated aquaculture was also the precursor to aquaponics (the integration of recirculating aquaculture and hydroponics) which, by the mid-2000s, was beginning to gain international momentum.

Gary Donaldson found the notion of aquaponics too limiting and he continued to promote a more holistic approach to small-scale food production through the inclusion of micro-livestock.

Aquaponics is, by definition, the combination of recirculating aquaculture and hydroponics for the production of fish and plants, however, Microponics suggests that recirculating aquaculture can be advantageously integrated with virtually any plant growing system.

The principal issue surrounding aquaponics is (like most aquaculture) its reliance on fishmeal and oil from wild catch marine species. The greater bio-diversity inherent in Microponics offers the opportunity to reduce or even eliminate this dependency.

Embodied within Microponics

The integrations embodied within Microponics include:

- Fish and crustaceans
- Vegetables, herbs and fruit
- Chickens for meat and eggs
- Japanese (Coturnix) Quail
- Rabbits and other microlivestock
- Muscovies

- Geese and other waterfowl
- Live animal protein - worms, Black Soldier fly larvae, mealworms
- Fodder plants - including duckweed
- Snails.

Given its emphasis on backyard food production, Microponics tends to focus on smaller micro-livestock species but, where space and local planning laws permit, the concept can be expanded to include traditional species of pigs, goats, sheep and even micro-cattle breeds like the Dexter.

Advantages

The advantages of Microponics food production systems include:

1. Empowerment of families to produce their own clean, fresh food.
2. Conservation through water reuse and recycling.
3. Organic fertilization of plants with natural fish emulsion.
4. Efficient use of space through Its small footprint -.
5. Reduced reliance on purchased livestock rations.
6. Its value as a learning resource for students of all ages.

Disadvantages

Some conceivable disadvantages with microponics are:

1. Initial expense for housing, tank, plumbing, pump/s, and grow beds.
2. The infinite number of ways in which a system can be configured lends itself to varying results, conflicting research, and successes or failures.
3. In heavily urbanised environments, a lack of space or local regulations might limit certain possible integrations.

Qian Hu Corporation

Qian Hu Corporation Limited is a Singapore-based ornamental fish service provider, with services ranging from the farming, importing, exporting and distribution of ornamental fish, to their speciality of breeding Dragon Fish (aka Arowana). Qian Hu exports more than 500 species and varieties of ornamental fish. Other services they offer also include the manufacturing and distribution of more than 5000 types of aquarium and pet accessories locally, as well as worldwide. Singapore exports one-third of the world's ornamental fish and Qian Hu has been a major contributor to this massive trade. It exports fish to more than 45 countries across the globe, with primary markets in Japan, China, Taiwan, the United Kingdom, Germany, France, as well as the rest of Europe and North Asia. Qian Hu has offices and outlets in three other countries: Malaysia,

China and Thailand, and hopes to eventually expand its reach even further across the Asian region.

History

The humble beginnings of Qian Hu go back to the 1980s when Yap Tik Huay, father of current Managing Director Kenny Yap, and Yap Tik Huay's brother, Yap Hey Cha ran a pig farm, rearing pigs as a family business.

During that time, however, the Singapore government was in the midst of making efforts to relocate and eventually wipe out all pig farms in the country not only because of the pollution they were causing, but also to make room for other urban and residential developments.

The Yap family's pig farm was no exception to the government's actions and in 1985, they had no choice but to convert their pig farm into a fish farm. By this time, three of Yap Tik Huay's sons, Yap Peng Heng, Yap Hock Huat and Yap Kim Choon, had begun working in the family business to lend a helping hand. The cubicles that originally penned in pigs were converted to concrete ponds for breeding guppies meant to be sold to local ornamental fish exporters.

In the Great Singapore Flood of 1989, the family's entire stock of over 4000 fish was washed away in a heavy thunderstorm. Left with nothing, the Yaps had to start anew, renaming their family business to hallmark their new beginning. The family business became known as Qian Hu (meaning 'Thousand Lakes' in Mandarin) and Kenny Yap, together with his two cousins Andy and Alvin Yap, joined the business in hopes of helping to pull it back up on its feet.

Thus, in 1990, Qian Hu ventured into wholesale, island-wide distribution and began to increase the range and variety of ornamental fish they sold. Aside from the guppies, Koi and Goldfish they sold at that time, fishes from the Chiclid (Discus included) and Gourami families were some of the new varieties introduced. Besides that, Qian Hu also began to manufacture its own accessories in an attempt to branch out its business.

Then, in 1992, Qian Hu eventually began importing ornamental fish and exporting them worldwide, making a name for itself as one of the world's leading exporters of ornamental fish. At the same time, Qian Hu took the opportunity to begin exporting its manufactured aquarium accessories.

On December 12, 1998, the company was incorporated under the Act as Qian Hu Fish Farm Pte Ltd to take over the partnership of Qian Hu Trading and Yi Hu, respectively established in 1988 and 1989.

Qian Hu Corporation Limited was listed on the Singapore Exchange's (SGX) Stock Exchange of Singapore Dealing and Automated Quotation system (SESDAQ) in the year 2000. By November 2002, it was moved to the Main Board of the Singapore Exchange.

Mission and Vision

Mission: Qian Hu's mission is "to create shareholders' value by becoming a world class ornamental fish and accessories company through innovative and quality products and services".

Vision

Qian Hu's vision is "to be the World's Biggest Ornamental Fish and Aquarium & Pet Accessories Service Provider".

Their objectives for this vision include:

- Becoming the world's number one ornamental fish exporter
- Escalating export of aquarium and pet accessories and to make "Ocean Free" one of the most recognised aquarium accessories brand
- Being the most innovative and profitable Dragon Fish breeder
- Being one of the top 3 manufacturers of aquarium accessories in China
- Expanding distribution capabilities from owning the business to owning the customers

Key People

Kenny Yap

Qian Hu's current Managing Director and Executive Chairman, Kenny Yap, is often cred for the Group's success, as his many innovative ideas have brought the company much acclaim.

The youngest of the seven sons in the Yap family, Yap was allowed the privilege of being educated overseas, unlike his elder brothers who had to remain in Singapore in order to help put at the farm. Yap graduated from Ohio State University with first class honours in Business Administration and returned home to the dwindling guppy farm. Seeing its condition, he, his brothers and two cousins banded together to revive the family business.

Since then, Yap has been working closely with his cousins and relatives to bring the company success. He was an active member of the Young Entrepreneur's Organisation and was conferred the Singapore Youth Award in 1998.

Branding

Qian Hu's corporate logo is that of a Chinese high fin banded shark (*Myxocyprinus asiaticus*), or what the company commonly refers to as a high-fin loach.

The story behind their use of the high-fin loach as their mascot comes not long after the Yap family faced the major setback of having lost their entire crop of guppies to the massive flood in 1989. In an attempt to get the business going again, Qian Hu invested its money in raising high-fin loaches. These fishes, at the time, cost SGD 100 to 200 each for a small specimen and the Yap family had been trying to bring the fish up to the desirable market size in order to sell them at a premium. However, tragedy struck whilst they expanded their premises. The move had caused a 'shocking wave' that travelled through the tanks of fish, subsequently killing all the high-fin loaches over a mere span of two to three days. This 'shocking wave', was in fact the vibrations given off by the power tools used to install new fish tanks. The family had not known that he fish happened to be highly sensitive to these vibrations. Once again, they were left with nothing but two grim lessons. As Kenny Yap recites time and again in his interviews, the lessons learnt were to not "put all your eggs in one basket" and to remember the importance of knowing a product well before investing in it. Hence, their mascot serves as a reminder to them of the mistakes they had made and to never make the same ones again.

Regional Expansion

Qian Hu has been distributing aquarium and pet accessories to China, Malaysia and Thailand since 1999.

In 2004, Qian Hu launched its retail store strategy, "Qian Hu – The Pet Family", all across Asia, with outlets in major cities like Shanghai, Nanjing, Kuala Lumpur, Batu Pahat, Bangkok and Bandung. Each of these retail chain stores sells ornamental fish as well as aquarium and pet accessories, some even providing grooming services for pets.

China

Qian Hu was involved in a joint-venture and later incorporated with The China Aquaculture Society to form Beijing Qian Hu Aquarium and Pets Co., Ltd (also known as Qian Yang) in 1993. The company, based in Beijing, supplied cold-water ornamental fish and aquarium accessories to the northern areas of China. Two years passed and Qian Hu had acquired full control and ownership of the joint-venture company in 1995.

Together with a Taiwanese partner, another joint-venture company was set up in Guangzhou, China, in 2001. Known as Wan Jiang Technology Co Ltd, the joint-venture company produces aquarium and pet accessories of the Group's propriety brands, as well as other third party brands and distributes them to Qian Hu's regional subsidiaries in other countries such as Japan, Germany and the United Kingdom. Wan Jiang eventually became a subsidiary of Qian Hu when it was decided that Qian Hu would increase its stake in Wan Jiang.

Malaysia

2002 saw the setting up of a new fish division of Qian Hu in Malaysia. In an attempt to ensure a reliable and stable source of Dragon Fish for Qian Hu's overseas subsidiaries and markets in Taiwan and Japan, the Group acquired one of the top Dragon Fish breeders in Batu Pahat, Malaysia in 2003—Kim Kang Aquaculture Sdn Bhd.

Thailand

Besides the joint-venture in China in 2001, Qian Hu also successfully set up a new subsidiary in Thailand that year. Thai Qian Hu Company Ltd, as it was later named, was 66% owned by Qian Hu in 2000. During that year, distribution and export of aquarium and pet accessories had to be put on hold for a period of time as Qian Hu awaited the approval of its license by Thailand's Ministry of Commerce (MOC) for the commencement of business in Thailand. The application of the license was part of the procedure as directed by Thailand's Foreign Business Act (FBA).

However, it was reported in the *Singapore Business Times* that Qian Hu failed to get the license approved. Thus the company attempted to restructure such that it could maintain control of the company but still operate in Thailand without a foreign business license.

When Qian Hu finally gained the approval of Thailand's Board of Investments to export and distribute ornamental tropical fish, in 2002 it entered a twenty joint venture with a pair of Thai nationals: Mrs. Somchit Purananant and Mr. Surajit Maneechote.

Awards and Achievements

Being awarded three ISO 9002s and an ISO 14001 were major achievements for Qian Hu as such awards were firsts for fish farms in Singapore and around the world.

The ISO 9002 certification is given to companies that have outstanding quality management systems. Qian Hu's first ISO 9002 was awarded to Qian Hu Trading in 1996 by SGS Yarsley for their conditioning and packing of their ornamental fish for export. The second came just a year after that, where Wan Hu division was awarded the ISO 9002 by Bureau Vertitas Quality International for their import and export, breeding, raising and quarantine procedures for the company's prided Dragon Fish. Then, in 2000, Yi Hu division was awarded the ISO 9002 certification by SGS Yarley for the retail and wholesale of aquarium and pet accessories locally and abroad.

Having learnt from the guppy stock disaster in 1989, Qian Hu had their 42-hectare premises landscaped to avoid the overflowing of ponds and tanks that held their valuable stocks of fish. Runoffs of rainwater are collected and recycled for use in the ponds, tanks and even toilets. This one-of-a-kind system that reuses and recycles water was what earned Qian Hu the ISO 14001 certification. The ISO 14001 is awarded to companies that have excellent environmental management systems. Some of Qian Hu's many titles, awards and achievements include being the "first small-cap company in Singapore to win a best managed board award" and two Singapore corporate awards: Best Investor Relationship and Gold Medal in Annual Report.

Fish Pond

A fish pond, or fishpond, is a controlled pond, artificial lake, or reservoir that is stocked with fish and is used in aquaculture for fish farming, or is used for recreational fishing or for ornamental purposes. In the medieval European era it was typical for monasteries and castles (small, partly self-sufficient communities) to have a fish pond.

History

During the winter, supplying fresh food for a castle garrison was a constant struggle. Although meat would be available from deer parks, this couldn't supply the needs of the whole household. A fish pond provided an elegant solution. As long as there was a natural flow of water into the pond, fish required no feeding and were available all year round. There would usually be a series of ponds, with fish being moved between them as they grew.

Aquaculture

Fish pond are used today in aquaculture. They are common in:

- Canada
- Europe, especially in the Czech Republic (Rožmberk Pond, Velké Dáøko, Máchovo jezero), where common carp may be kept.
- The Philippines where Milkfish, Tilapia, crabs, lobsters, Tiger shrimp, snails and others may be kept.
- East Asia, especially in Japan with koi.

Fish ponds are also being promoted in developing countries. They provide a source of food and income from the sale of fish for small farmers and can also supply irrigation needs and water for livestock.

Koi Pond

Koi ponds are ponds used for holding koi, usually as part of a landscape. Koi ponds can be designed specifically to promote health and growth of the Nishkigoi or Japanese Ornamental Carp. The architecture of the koi pond can have a great effect on the health and wellbeing of the koi. The practice of keeping koi often revolves around "finishing" a koi at the right time. The concept of finishing means that the fish has reached is highest potential. Koi clubs hold shows where koi keepers bring their fish for judging.

Figure: Ornamental pond stocked with koi

Components

Skimmer

The skimmer allows water to be drawn from the surface of the pond. It collects leaves, pollen, twigs, uneaten food and all other kinds of floating debris. The skimmer usually has a clean out basket that can be quickly emptied on a regular basis to allow the skimmer to run properly. Most floating skimmers will also have a foam that sits underneath the basket to filter out the finer particles.

Bottom Drain

Bottom drains are not required in water gardens but are very beneficial for Koi Ponds. When used in a pond that does not have rocks on the bottom, a bottom drain allows the heavy solids to be carried to the mechanical filter.

Mechanical Filter

Mechanical filtration can be accomplished many different ways. The job of this filter is to trap solids, preventing them from clogging the Biological filter. The mechanical filter should be back washed or cleaned out often. Types of mechanical filters include Vortex, brushes, matting, sand and gravel, sieve screen, and settlement chamber.

Biological Filter

Biological filters convert the nitrogenous wastes from the fish. This cycle is called the nitrogen cycle. A biofilter can be constructed many different ways. It is important for the koi keeper to understand how the filter is to be cleaned before they install one. Proper and regular cleaning of the mechanical and biological filters is critical for the health and quality of the koi. Bio-filters are sometimes divided into sub groups such as aerated or non-aerated. Types of bio-filters include:

- Moving bed filter - aerated
- Bakki shower or Trickle filter - aerated

- Sand filter - not aerated
- Cross-flow filter - typically not aerated, the ERIC filter is aerated.
- Bead filter

Ultraviolet Light

An ultraviolet light is used to make algae flocculate (form into clumps), so that they can be removed by mechanical filtration.

Water and Air Pumps

Water pumps move water through the filter system and back to the pond in a recirculating manner. The important thing to understand about pumps is that they be sized to the pond and the filter system. When the total back pressure in the system is considered, a pump should be circulating the total volume of water at least once per hour for proper water quality. A pond air pump can be used to increase dissolved oxygen.

4

Water Garden

Water gardens, also known as aquatic gardens, are a type of man-made water feature. A water garden is defined as any interior or exterior landscape or architectural element whose primarily purpose is to house, display, or propagate a particular species or variety of aquatic plant.

Although a water garden's primary focus is on plants, they will sometimes also house ornamental fish, in which case the feature will be a fish pond.

Although water gardens can be almost any size or depth, they are typically small and relatively shallow, generally less than twenty inches in depth. This is because most aquatic plants are depth sensitive and require a specific water depth in order to thrive. The particular species inhabiting each water garden will ultimately determine the actual surface area and depth required.

When the aquatic flora and fauna is balanced, an aquatic ecosystem is created that supports sustainable water quality/water clarity.

Elements such as: fountains, statuary, waterfalls, boulders, underwater lighting, lining treatments, edging detailing, in-water and bankside planting and watercourses can be combined with the pool to add visual interest and integration with the local landscape and environment.

Water gardens, and all water features in general have been a part of public and private gardens since ancient Persian gardens and Chinese gardens. Water features have been present and well represented in every era and culture that has included gardens in their landscape and architectural environments. Up until the rise of the industrial age, that introduced the modern water pump, water was not recirculated, but was diverted from rivers and springs into the water garden, exiting into agricultural fields or natural watercourses. Water features were historically used for plant and fish production for food purposes as well as ornamental aesthetics.

Though a water garden is restricted to a particular type of natural or man-made water feature, used for a relatively specific purpose or intended use, there are many other types of water feature types, styles and designs.

Types of water features

- Natural Water Feature
- Man-made Water Feature
- Naturalistic Water Feature
- Disappearing Water Feature
- Live Water Feature
- Sterile Water Feature
- Water Garden
- Bog Garden
- Rain Garden/Bio Retention System/Rain Harvesting
- Aquatic Container Garden
- Pool/Shallow Pool/Tide Pool
- Reflection Pool/Reflecting Pool
- Formal Pool
- Swimming Pool
- Pond/Fish Pond/Backyard Pond/Garden Pond
- Naturalistic pond
- Wildlife Pond/Habitat Pond
- Koi Pond
- Swimming Pond
- Water Courses
- Brooks
- Creeks
- Streams
- Rivers

- Runnel
- Rile
- Weeping Wall
- Water Wall
- Fountain/Formal Fountain
- Disappearing Fountain
- Tabletop Fountain
- Wall Fountain
- Spitter Fountain
- Bubbler Fountain
- Floating Fountain
- Water Falls
- Hydroponics
- Lotus pool
- Rice paddy
- Riparian zone restoration
- Wildlife garden - *with water-source component.*
- Stream pool
- Plunge pool
- Plunge basin
- Spring (hydrology)
- Seep (hydrology)
- Bogs
- Wetlands
- Mangrove swamp habitat
- Wild River
- Lakes
- Halka lever

Water Follies

In the sixteenth century, Europe had a renewed interest in Greek thought and philosophy including the works of Hero of Alexandria about hydraulics and pneumatics. His devices, such as temple doors operated by invisible weights or flowing liquids, and mechanical singing birds powered by steam, motivated several

European palaces to create similar clever devices to enhance their public image. In Italy several royal houses constructed large water gardens incorporating mechanical devices in water settings.

The best-known is the Villa d'Este at Tivoli, constructed in 1550 AD. Popularity spread across Europe with the well known water garden at Hellbrunn Palace built, with many water-powered human and animal performing figures and puppet theaters, and folly fountains that erupted without notice to surprise one.

Stream Gardens

On a constructed stream, placing rocks in the path of the water makes small patterns, rapids and waterfalls. The rocks disrupt the waterflow, causing splashing and bubbles to form, which can make pleasant sounds and micro-habitats for plants , fish, and wildlife. Well placed rocks can stimulate splashing water that adds oxygen to prevent hypoxia, with the more bubbles - the more dissolved oxygen in the water.

Aquatic Flora

Submerged plants are those that live almost completely under the water, sometimes with leaves or flowers that grow to the surface such as with the water lily. These plants are placed in a pond or container usually 1–2 ft (0.30–0.61 m) below the water surface. Some of these plants are called oxygenators because they create oxygen for the fish that live in a pond. Examples of submerged plants are:

- Water lily (Hardy and Tropical)
- Hornwort (*Ceratophyllum demersum*)
- Marginal plants are those that live with their roots under the water but the rest of the plant above the surface. These are usually placed so that the top of the pot is at or barely below the water level. Examples of these are:
- Iris or Flag (*Iris* spp.)
- Water-crowfoot (*Ranunculus fluitans*)
- Bulrush (*Scirpus lacustris*)
- Cattail (*Typha latifolia*)
- Taro (*Colocasia esculenta*)
- Arrowhead (*Sagittaria latifolia*)
- Lotus (*Nelumbo* spp.)
- Pickerelweed (*Pontederia cordata*)
- Floating plants are those that are not anchored to the soil at all, but are free-floating on the surface. In water gardening, these are often used as

a provider of shade to reduce algae growth in a pond. These are often extremely fast growing/multiplying. Examples of these are:

- Mosquito ferns (*Azolla* spp.)
- Water-spangle (*Salvinia* spp.)
- Water-clover (*Marsilea vestita*)
- Water Lettuce (*Pistia stratiotes*)
- Water Hyacinth (*Eichhornia crassipes*)

Some areas of the United States do not allow certain of these plants to be sold or kept as they have become invasive species in warmer areas of the country, such as Florida and California.

Algae

Algae are found in all ponds. There are hundreds of species of algae that can grow in garden ponds but they are only usually noticed when they become abundant. Algae often grow in very high densities in ponds because of the high nutrient levels that are typical of garden ponds. Generally alga attaches itself to the sides of the pond and remains innocuous.

Some species of algae, namely the dreaded 'blanket weed' can grow up to a foot a day under ideal conditions and can rapidly clog a garden pond. On the other hand, free floating algae are microscopic and are what cause pond water to appear green.

Fauna

Fish

Often the reason for having a pond in a garden is to keep fish, often koi, though many people keep goldfish. Both are hardy, colourful fish which require no special heating, provided the pond is located in an area which does not have extremes of temperature that would affect the fish. If fish are kept, pumps and filtration devices are usually needed in order to keep enough oxygen in the water to support them. In winter, a small heater may need to be used in cold climates to keep the water from freezing solid. Examples of common pond fish include:

- Ricefish (Himedaka)
- Mosquitofish
- Rosy Red Minnows
- White Cloud Mountain minnow
- Goldfish (Common, Comet, Shubunkin varieties, Wakin and the Fantail varieties. With the possible exception of some of the fantail varieties, the fancy goldfish are not suited to pond life.)

- Crucian carp
- Koi (Nishikigoi, Butterfly Koi and Ghost Koi)
- Mirror carp
- Carp
- Weather loach
- Golden Orfe
- Golden Tench
- Eel
- Catfish
- Bluegill
- Black bass
- Snakehead
- Goby

Crustacean:

- Crayfish
- Freshwater Prawn

Snails:

- River snail.

Small aquatic snails are usually in ponds which have plants. Some people purchase apple snails to keep in their water garden. Another common variety is the Melantho snail.

Herpetofauna

Ponds located in suburban and rural areas often attract amphibians such as Common Frogs and Fire Salamanders, and reptiles such as turtles, lizards, and snakes.

Bird:

- Wild duck
- Domestic duck.

Predators

Garden ponds can attract attention from predators such as (in North America) raccoons, Herons, snakes, and domestic cats.

These predators can be a danger to fish. Owners of koi are often particularly careful to create protected areas as some varieties are very expensive.

Goldfish

The goldfish (*Carassius auratus auratus*) is a freshwater fish in the family Cyprinidae of order Cypriniformes. It was one of the earliest fish to be domesticated, and is one of the most commonly kept aquarium fish.

A relatively small member of the carp family (which also includes the koi carp and the crucian carp), the goldfish is a domesticated version of a less-colourful carp (*Carassius auratus*) native to east Asia. It was first domesticated in China more than a thousand years ago, and several distinct breeds have since been developed. Goldfish breeds vary greatly in size, body shape, fin configuration and colouration (various combinations of white, yellow, orange, red, brown, and black are known).

History

Starting in ancient China, various species of carp (collectively known as Asian carps) have been domesticated and reared as food fish for thousands of years. Some of these normally gray or silver species have a tendency to produce red, orange or yellow colour mutations; this was first recorded in the Jin Dynasty (265–420).

Figure: A western aquarium of the 1850s of the type that contained goldfish among other coldwater species

During the Tang Dynasty (618–907), it was popular to raise carp in ornamental ponds and watergardens. A natural genetic mutation produced gold (actually yellowish orange) rather than silver colouration. People began to breed the gold variety instead of the silver variety, keeping them in ponds or other bodies of water. On special occasions at which guests were expected they would be moved to a much smaller container for display.

By the Song Dynasty (960–1279), the domestication of goldfish was firmly established. In 1162, the empress of the Song Dynastyordered the construction of a pond to collect the red and gold variety. By this time, people outside the imperial family were forbidden to keep goldfish of the gold (yellow) variety, yellow being the imperial colour. This is probably the reason why there are more orange goldfish than yellow goldfish, even though the latter are genetically easier to breed.

During the Ming Dynasty (1368-1644), goldfish also began to be raised indoors, which led to the selection for mutations that would not be able to survive in ponds. The occurrence of other colours (apart from red and gold) was first recorded in 1276. The first occurrence of fancy-tailed goldfish was recorded in the Ming Dynasty. In 1603, goldfish were introduced to Japan, where theRyukin and Tosakin varieties were developed. In 1611, goldfish were introduced to Portugal and from there to other parts ofEurope.

During the 1620s, goldfish were highly regarded in southern Europe because of their metallic scales, and symbolised good luck and fortune. It became tradition for married men to give their wives a goldfish on their one-year anniversary, as a symbol for the prosperous years to come. This tradition quickly died, as goldfish became more available, losing their status. Goldfish were first introduced to North America around 1850 and quickly became popular in the United States.

Related Species

Figure: A wild Prussian carp (Carassius auratus gibelio)

Figure: A Crucian carp (Carassius carassius)

Goldfish were bred from Prussian carp (*Carassius auratus gibelio*) in China, and they remain the closest wild relative of the goldfish. Previously, some sources claimed the Crucian carp (*Carassius carassius*) as the wild version of the goldfish. However, they are differentiated by several characteristics. *C. auratus* have a more pointed snout while the snout of a *C. carassius* is well rounded. *C. gibelio* often has a grey/greenish colour, while crucian carps are always golden bronze.

Juvenile crucian carp have a black spot on the base of the tail which disappears with age. In C. auratus this tail spot is never present. *C. auratus* have fewer than 31 scales along the lateral line while crucian carp have 33 scales or more. When found in nature, goldfish are olive green. Introduction of goldfish into the wild can cause problems for native species. Goldfish can hybridise with certain other species of carp. Within three breeding generations, the vast majority of the hybrid spawn revert to their natural olive colour. The mutation that gave rise to the domestic goldfish is also known from other cyprinid species, such as common carp and tench. Koi may also interbreed with the goldfish to produce a sterile hybrid fish.

There are many different varieties of domesticated goldfish. Fancy goldfish are unlikely to survive in the wild because of their bright fin colours; however the hardier varieties such as the Shubunkin may survive long enough to breed with wild cousins. Common and comet goldfish can survive, and even thrive, in any climate that can support a pond.

Varieties of Domesticated Goldfish

Selective breeding over centuries has produced several colour variations, some of them far removed from the "golden" colour of the originally domesticated fish. There are also different body shapes, fin and eye configurations. Some extreme versions of the goldfish live only in aquariums—they are much less hardy than varieties closer to the "wild" original. However, some variations are hardier, such as the Shubunkin. Currently, there are about 300 breeds recognised in China. The vast majority of goldfish breeds today originated from China. Some of the main varieties are:

Chinese Goldfish Classification: Chinese tradition classifies goldfish into four main types. These classifications are not commonly used in the West.

- Ce (may also be called "grass")—Goldfish without fancy anatomical features. These include the common goldfish, comet goldfish and Shubunkin.
- Wen—Goldfish have a fancy tail, e.g., Fantails and Veiltails ("Wen" is also the name of the characteristic headgrowth on such strains as Oranda and Lionhead)
- Dragon Eye—Goldfish have extended eyes, e.g., Black Moor, Bubble Eye, and Telescope Eye
- Egg—Goldfish have no dorsal fin, and usually have an 'egg-shaped' body, e.g., Lionhead (note that a Bubble Eye without a dorsal fin belongs to this group)

Size

As of April 2008, the largest goldfish in the world was believed by the BBC to measure 19 inches (48 cm), and be living in the Netherlands. At the time, a goldfish named "Goldie", kept as a pet in a tank in Folkestone, England, was measured as 15 inches (38 cm) and over 2 pounds (0.91 kg), and named as the second largest in the world behind the Netherlands fish. The secretary of the Federation of British Aquatic Societies (FBAS) stated of Goldie's size that "I would think there are probably a few bigger goldfish that people don't think of as record holders, perhaps in ornamental lakes". In July 2010, a goldfish measuring 16 inches (41 cm) and 5 pounds (2.3 kg) was caught in a pond in Poole, England, thought to have been abandoned there after outgrowing a tank.

In Ponds

Goldfish are popular pond fish, since they are small, inexpensive, colourful, and very hardy. In an outdoor pond or water garden, they may even survive for brief periods if ice forms on the surface, as long as there is enough oxygen remaining in the water and the pond does not freeze solid.

Common goldfish, London and Bristol shubunkins, jikin, wakin, comet and some hardier fantail goldfish can be kept in a pond all year round in temperate and subtropical climates. Moor, veiltail, oranda and lionhead can be kept safely in outdoor ponds only in the summer, and in more tropical climates.

Small to large ponds are fine though the depth should be at least 80 centimetres (31 in) to avoid freezing. During winter, goldfish become sluggish, stop eating, and often stay on the bottom of the pond.

This is completely normal; they become active again in the spring. A filter is important to clear waste and keep the pond clean. Plants are essential as they act as part of the filtration system, as well as a food source for the fish. Plants are further beneficial since they raise oxygen levels in the water.

Compatible fish include rudd, tench, orfe and koi, but the latter require specialised care. Ramshorn snails are helpful by eating any algae that grows in the pond. Without some form of animal population control, goldfish ponds can easily become overstocked. Fish such as orfe consume goldfish eggs.

In Aquaria

Like most carp, goldfish produce a large amount of waste both in their faeces and through their gills, releasing harmful chemicals into the water. Build-up of this waste to toxiclevels can occur in a relatively short period of time, and can easily cause a goldfish's death. For common and comet varieties, each goldfish should have about 20 US gallons (76 l; 17 imp gal) of water. Fancy goldfish (which are smaller) should have about 10 US gallons (38 l; 8.3 imp gal) per goldfish. The water surface area determines how muchoxygen diffuses and dissolves into the water. A general rule is have 1 square foot (0.093 m^2). Active aeration by way of a water pump, filter or fountain effectively increases the surface area.

The goldfish is classified as a coldwater fish, and can live in unheated aquaria at a temperature comfortable for humans. However, rapid changes in temperature (for example in an office building in winter when the heat is turned off at night) can kill them, especially if the tank is small. Care must also be taken when adding water, as the new water may be of a different temperature. Temperatures under about 10 °C (50 °F) are dangerous to fancy varieties, though commons and comets can survive slightly lower temperatures. Extremely high temperatures (over 30 °C (86 °F) can also harm goldfish. However, higher temperatures may help fight protozoan infestations by accelerating the parasite's life-cycle—thus eliminating it more quickly. The optimum temperature for goldfish is between 20 °C (68 °F) and 22 °C (72 °F).

Like all fish, goldfish do not like to be petted. In fact, touching a goldfish can endanger its health, because it can cause the protective slime coat to be damaged or removed, exposing the fish's skin to infection from bacteria or water-born parasites. However, goldfish respond to people by surfacing at feeding time, and can be trained or acclimated to taking pellets or flakes from human fingers. The reputation of goldfish dying quickly is often due to poor care. The lifespan of goldfish in captivity can extend beyond 10 years.

If left in the dark for a period of time, goldfish gradually change colour until they are almost gray. Goldfish produce pigment in response to light, in a similar manner to how human skin becomes tanned in the sun. Fish have cells called chromatophores that produce pigments which reflect light, and give the fish colouration. The colour of a goldfish is determined by which pigments are in the cells, how many pigment molecules there are, and whether the pigment is grouped inside the cell or is spaced throughout the cytoplasm.

Because goldfish eat live plants, their presence in a planted aquarium can be problematic. Only a few aquarium plant species for example *Cryptocoryne* and *Anubias*, can survive around goldfish, but they require special attention so that they are not uprooted. Plastic plants are often more durable, but the branches can irritate or harm a fish that touches one.

Feeding

In the wild, the diet of goldfish consists of crustaceans, insects, and various plant matter. Like most fish, they are opportunistic feeders and do not stop eating on their own accord. Overfeeding can be deleterious to their health, typically by blocking the intestines. This happens most often with selectively bred goldfish, which have a convoluted intestinal tract. When excess food is available, they produce more waste and faeces, partly due to incomplete protein digestion. Overfeeding can sometimes be diagnosed by observing faeces trailing from the fish's cloaca.

Goldfish-specific food has less protein and more carbohydrate than conventional fish food. It is sold in two consistencies—flakes that float, and pellets that sink. Enthusiasts may supplement this diet with shelled peas (with outer skins removed), blanched green leafy vegetables, and bloodworms. Young goldfish benefit from the addition of brine shrimp to their diet. As with all animals, goldfish preferences vary.

Behaviour

Behaviour can vary widely both because goldfish live in a variety of environments, and because their behaviour can be conditioned by their owners.

Goldfish have strong associative learning abilities, as well as social learning skills. In addition, their visual acuity allows them to distinguish between individual humans. Owners may notice that fish react favourably to them (swimming to the front of the glass, swimming rapidly around the tank, and going to the surface mouthing for food) while hiding when other people approach the tank. Over time, goldfish learn to associate their owners and other humans with food, often "begging" for food whenever their owners approach. Responses from a blind goldfish proved that it recognised one particular family member and a friend by voice, or vibration of sound. This behaviour was remarkable because it showed that the fish recognised the vocal vibration or sound of two people specifically out of seven in the house.

Goldfish are gregarious, displaying schooling behaviour, as well as displaying the same types of feeding behaviours. Goldfish may display similar behaviours when responding to their reflections in a mirror.

Goldfish that have constant visual contact with humans also stop considering them to be a threat. After being kept in a tank for several weeks, sometimes months, it becomes possible to feed a goldfish by hand without it shying away.

Goldfish have learned behaviours, both as groups and as individuals, that stem from native carp behaviour. They are a generalist species with varied feeding, breeding, and predator avoidance behaviours that contribute to their success. As fish they can be described as "friendly" towards each other. Very rarely does a goldfish harm another goldfish, nor do the males harm the females during breeding. The only real threat that goldfish present *to each other* is competing for food. Commons, comets, and other faster varieties can easily eat all the food during a feeding before fancy varieties can reach it. This can lead to stunted growth or possible starvation of fancier varieties when they are kept in a pond with their single-tailed brethren. As a result, care should be taken to combine only breeds with similar body type and swim characteristics.

Intelligence

Goldfish have a memory-span of at least three months and can distinguish between different shapes, colours and sounds. Goldfish vision is among the most studied of all vision in fishes. By using positive reinforcement, goldfish can be trained to recognise and to react to light signals of different colours or to perform tricks. Fish respond to certain colours most evidently in relation to feeding. Fish learn to anticipate feedings provided they occur at around the same time everyday.

Reproduction

Goldfish may only grow to sexual maturity with enough water and the right nutrition. Most goldfish breed in captivity, particularly in pond settings. Breeding usually happens after a significant temperature change, often in spring. Males chase gravid female goldfish (females carrying eggs), and prompt them to release their eggs by bumping and nudging them.

Goldfish, like all cyprinids, are egg-layers. Their eggs are adhesive and attach to aquatic vegetation, typically dense plants such as *Cabomba* or *Elodea* or a spawning mop. The eggs hatch within 48 to 72 hours.

Within a week or so, the fry begins to assume its final shape, although a year may pass before they develop a mature goldfish colour; until then they are a metallic brown like their wild ancestors. In their first weeks of life, the fry grow quickly—an adaptation born of the high risk of getting devoured by the adult goldfish (or other fish and insects) in their environment.

Some highly bred goldfish can no longer breed naturally due to their altered shape. The artificial breeding method called "hand stripping" can assist nature, but can harm the fish if not done correctly. In captivity, adults may also eat young that they encounter.

Mosquito Control

Like some other popular aquarium fish, such as the guppy, goldfish and other carp are frequently added to stagnant bodies of water to reduce mosquito populations. They are used to prevent the spread of West Nile Virus, which relies on mosquitoes to migrate. However, introducing goldfish has often had negative consequences for local ecosystems.

Controversy Over Proper Treatment

Some countries ban the sale of traditional fishbowls under animal welfare legislation due to the risk of stunting, deoxygenation and ammonia/nitrite poisoning in such a small environment. Because of their large oxygen needs and high waste output, such bowls are no longer considered appropriate housing for goldfish.

In many countries, carnival and fair operators commonly give goldfish away in plastic bags as prizes. In late 2005 Rome banned the use of goldfish and other animals as carnival prizes. Rome has also banned the use of "goldfish bowls", on animal cruelty grounds. In the United Kingdom, the government proposed banning this practice as part of its Animal Welfare Bill, though this has since been amended to only prevent goldfish being given as prizes to unaccompanied minors. In Japan, during summer festivals and religious holidays (ennichi), a traditional game called goldfish scooping is played, in which a player scoops goldfish from a basin with a special scooper. Sometimes bouncy balls are substituted for goldfish.

Although edible, goldfish are rarely eaten. A fad among American college students for many years was swallowing goldfish as a stunt and as a fraternity initiation process. The first recorded instance was in 1939 at Harvard University. The practice gradually fell out of popularity over the course of several decades and is rarely practiced today.

Cyprinidae

The family Cyprinidae, from the Ancient Greek *kyprînos* ("carp"), consists of the carps, the true minnows, and their relatives (for example, the barbs and barbels). Commonly called the carp family or the minnow family, its members are also known as cyprinids. It is the largest family of fresh-water fish, with over 2,400 species in about 220 genera. The family belongs to the order Cypriniformes, of whose genera and species the cyprinids make up two-thirds.

Description

Figure: Giant Barbs (*Catlocarpio siamensis*) are the largest members of this family

Cyprinids are stomachless fish with toothless jaws. Even so, food can be effectively chewed by the gill rakers of the specialised last gill bow. These pharyngeal teeth allow the fish to make chewing motions against a chewing plate formed by a procession of the skull. The pharyngheal teeth are species specific and are used by specialists to determine the species. Strong pharyncheal teeth allow fish like the common carp and ide to eat hard baits like snails and bivalves.

Hearing is a well-developed sense, since the cyprinds have the Weberian organ, three specialised vertebra processions that transfer motion of the gas bladder to the inner ear. This construction is also used to observe motion of the gas bladder due to atmospheric conditions or depth changes. The cyprinids are physostomes because the pneumatic duct is retained in adult stages and the fish are able to gulp air to fill the gas bladder or they can dispose excess gas to the gut.

The fish in this family are native to North America, Africa, and Eurasia. The largest cyprinid in this family is the Giant Barb (Catlocarpio *siamensis*), which may grow up to 3 metres (9.8 ft). The largest North American species is the Colorado Pikeminnow (*Ptychocheilus lucius*), of which individuals up to 6 feet (1.8 m) long and weighing over 100 pounds (45 kg) have been recorded.

On the other hand, many species are smaller than 5 centimetres (2.0 in). As of 2008, the smallest known freshwater fish is a cypriniform, *Danionella translucida*, reaching 12 millimetres (0.47 in) at the longest. All fish in this family are egg-layers and most do not guard their eggs, however, there are a few species that build nests and/or guard the eggs. The bitterling-like cyprinids (Acheilognathinae) are notable for depositing their eggs in bivalve molluscs, where the young grow up until able to fend for themselves.

Most cyprinids feed mainly on invertebrates and vegetation probably due to the lack of teeth and stomach, but some species like the Asp specialise in fish. Many species ide, common rudd will eat small fish however when reaching a certain size. Even small species like the moderlieschen eat larvae of the common frog in artificial circumstances.

Some fish, such as the grass carp, are specialised in eating vegetation, some, such as the common nase, eat algae from hard surfaces, some, such as the black carp, specialise in snails, and some, such as the silver carp, are specialised filter feeders. For this reason, they are often introduced as a management tool to control various factors in the aquatic environment, such as aquatic vegetation and diseases transmitted by snails.

Relationship with Humans

Cyprinids are highly important food fish; they are fished and farmed across Eurasia. In land-locked countries in particular, cyprinids are often the major species of fish eaten because they make the largest part of biomass in most water types except for fast flowing rivers. In non-landlocked countries they are not very much appreciated due to the high number of bones. In Eastern Europe they are often prepared with traditional methods like drying and salting. The prevalence of inexpensive frozen fish products made this less important now than it was in earlier times. Nonetheless, in certain places they remain popular for food as well as recreational fishing, and have been deliberately stocked in ponds and lakes for centuries for this reason.

Cyprinids are popular for angling especially for match fishing (due to their dominance in biomass and numbers) and fishing for common carp because of its size and strength.

Several cyprinids have been introduced to waters outside their natural range to provide food, sport, or biological control for some pest species. The Common

Carp (*Cyprinus carpio*) and the Grass Carp (Ctenopharyngodon *idella*) are the most important of these, for example in Florida. In some cases, these have become invasive species that compete with native fishes or disrupt the environment. Carp in particular can stir up sediment, reducing the clarity of the water and making it difficult for plants to grow.

Numerous cyprinids have become important in the aquarium hobby, most famously the Goldfish, which was bred in China from the Prussian Carp (*Carassius (auratus) gibelio*). First imported into Europe around 1728, it was much fancied by Chinese nobility as early as 1150 AD and after it arrived there in 1502, also in Japan. In the latter country, from the 18th century onwards the Common Carp was bred into the ornamental variety known as koi – or more accurately *nishikigoi* , as *koi* simply means "Common Carp" in Japanese.

Other popular aquarium cyprinids include danionins, rasborines and true barbs. Larger species are bred by the thousands in outdoor ponds, particularly in Southeast Asia, and trade in these aquarium fishes is of considerable commercial importance. The small rasborines and danionines are perhaps only rivalled by characids and poecilid livebearers in their popularity for community aquaria.

One particular species of these small and undemanding danionines is the Zebrafish (*Danio rerio*). It has become the standard model species for studying developmental genetics of vertebrates, in particular fish.

Habitat destruction and other causes have reduced the wild stocks of several cyprinids to dangerously low levels; some are already entirely extinct. In particular, Leuciscinae from southwestern North America have been hit hard by pollution and unsustainable water use in the early-mid 20th century; most globally extinct Cypriniformes species are in fact Leuciscinae from the southwestern United States and northern Mexico.

Systematics

The massive diversity of cyprinids has so far made it difficult to resolve their phylogeny in sufficient detail to make assignment to subfamilies more than tentative in many cases. It is obvious that some distinct lineages exist – for example, Cultrinae and Leuciscinae, regardless of their exact delimitation, are rather close relatives and stand apart from Cyprininae –, but the overall systematics and taxonomy of the Cyprinidae remain a subject of considerable debate. A large number of genera are *incertae sedis*, too equivocal in their traits and/or too little-studied to permit assignment to a particular subfamily with any certainty.

Part of the solution seems that the delicate rasborines are the core group, consisting of minor lineages that have not shifted far from their evolutionary niche, or have co-evolved, for millions of years. These are among the most basal lineages of living cyprinids. Other "rasborines" are apparently distributed across the diverse lineages of the family. The validity and circumscription of proposed subfamilies like Labeoninae or Squaliobarbinae also remains doubtful, although the latter do appear to correspond to a distinct lineage. The sometimes-seen grouping of the large-headed carps (Hypophthalmichthyinae) with *Xenocypris*, on the other hand, seems quite in error. More likely, the latter are part of the Cultrinae.

The entirely paraphyletic "Barbinae" and the disputed Labeoninae might be better treated as part of the Cyprininae, forming a close-knit group whose internal relationships are still little known. The small African "barbs" do not belong in *Barbus* sensu stricto – indeed, they are as distant from the typical barbels and the typical carps (*Cyprinus*) as these are from *Garra* (which is placed in the Labeoninae by most who accept the latter as distinct) and thus might form another as of yet unnamed subfamily. However, as noted above, how various minor lineages tie into this has not yet been resolved; therefore such a radical move, though reasonable, is probably premature.

The Tench (*Tinca tinca*), a significant food species farmed in western Eurasia in large numbers, is unusual. It is most often grouped with the Leuciscinae, but even when these were rather loosely circumscribed, it always stood apart. A cladistic analysis of DNA sequence data of the S7 ribosomal protein intron 1 supports the view that it is distinct enough to constitute a monotypic subfamily. It also suggests that it may be closer to the small East Asian *Aphyocypris*, *Hemigrammocypris*, and *Yaoshanicus*. They would have diverged roughly at the same time from cyprinids of east-central Asia, perhaps as a result of the Alpide orogeny that vastly changed the topography of that region in the late Paleogene, when their divergence presumably occurred.

Koi or more specifically *nishikigoi* , are ornamental varieties of domesticated common carp (*Cyprinus carpio*) that are kept for decorative purposes in outdoor koi ponds or water gardens.

Koi varieties are distinguished by colouration, patterning, and scalation. Some of the major colours are white, black, red, yellow, blue, and cream. The most popular category of koi is the *Gosanke*, which is made up of the *Kohaku*, *Taisho Sanshoku*, and *Showa Sanshoku* varieties.

History

Cyprinus carpio or the common carp is a species of fish from the family Cyprinidae. The origins of the common carp trace to the Caspian Sea, where the fish naturally migrated to the Black and Aral Seas, east to eastern mainland Asia and west as far as the Danube River. Within the same family, the Prussian carp (*Carassius gibelio*) were first bred for colour mutations in China more than a thousand years ago, where selective breeding of led to the development of the goldfish. The first known selective breeding resulting in ornamental fish. The common carp was known to have been aquacultured in Europe by the Roman Empire, which could have spanned a time period of 27 BC to 400 AD. The common carpwas also aquacultured as a food fish at least as far back as the fifth century in China.

Common carp were first bred for colour in Japan in the 1820s, initially in the town of Ojiya in the Niigata prefecture on the northeastern coast of Honshu island. By the 20th century, a number of colour patterns had been established, most notably the red-and-white *Kohaku*. The outside world was not aware of the development of colour variations in koi until 1914, when the Niigata koi were exhibited in the annual exposition in Tokyo. At that point, interest in koi exploded throughout Japan. The hobby of keeping koi eventually spread worldwide. They are now commonly sold in most pet stores, with higher-quality fish available from specialist dealers.

Extensive hybridisation between different populations has muddled the historical zoogeography of the common carp. However, scientific consensus is that there are at least two subspecies of the common carp, one from Western Eurasia (*Cyprinus carpio carpio*) and another from East Asia (*Cyprinus carpio haematopterus*). One recent study on the mitochondrial DNA of various common carp indicate that koi are of the East Asian subspecies. However, another recent study on the mitochondrial DNA of koi have found that koi are descended from multiple lineages of common carp from both Western Eurasian and East Asian varieties. This could be the result of koi being bred from a mix of East Asian and Western Eurasian carp varieties, or being bred exclusively from East Asian varieties and being subsequently hybridised with Western Eurasian varieties (the butterfly koi is one known product of such a cross). Which is true has not been resolved.

Etymology

The word *koi* comes from Japanese, simply meaning "carp". It includes both the dull grey fish and the brightly coloured varieties. What are known as koi in English are referred to more specifically as *nishikigoi* in Japan (literally meaning "brocaded carp"). In Japanese, *koi* is a homophone for another word that means "affection" or "love"; koi are, therefore, symbols of love and friendship in Japan. An example of this can be seen in the short story by Mukôda Kuniko, "Koi-san". The koi is also an often recurring symbol in Irezumi, the Japanese art of traditional tattooing.

Varieties

Koi varieties are distinguished by colouration, patterning, and scalation. Some of the major colours are white, black, red, yellow, blue, and cream. While the possible colours are virtually limitless, breeders have identified and named a number of specific categories. The most popular category is **Gosanke**, which is made up of the *Kohaku*, *Taisho Sanshoku*, and *Showa Sanshoku* varieties. New koi varieties are still being actively developed. Ghost koi developed in the 1980s have become very popular in the United Kingdom; they are a hybrid of wild carp and Ogon koi, and are distinguished by their metallic scales. Butterfly koi (also known as longfin koi, or dragon carp), also developed in the 1980s, are notable for their long and flowing fins. They are hybrids of koi with Asian carp. Butterfly koi and ghost koi are considered by some to be not true *nishikigoi*.

The major named varieties include:

- *Kôhaku* is a white-skinned koi, with large red markings on the top. The name means "red and white"; *kohaku* was the first ornamental variety to be established in Japan (late 19th century).
- *Taishô Sanshoku* (or *Taisho Sanke)* is very similar to the *kohaku*, except for the addition of small black markings called *sumi* . This variety was first exhibited in 1914 by the koi breeder Gonzo Hiroi, during the reign of the Taisho Emperor. In America, the name is often abbreviated to just "Sanke". The kanji, Nr,, may be read as either *sanshoku* or as *sanke.*
- *Shôwa Sanshoku* (or *Showa Sanke*) is a black koi with red and white markings. The first *Showa Sanke* was exhibited in 1927, during the reign of the Showa Emperor. In America, the name is often abbreviated to just "Showa". The amount of *shiroji* on *Showa Sanke* has increased in modern times, to the point that it can be difficult to distinguish from *Taisho Sanke.* The kanji, Nr,, may be read as either *sanshoku* or as *sanke.*
- *Tanchô* is any koi with a solitary red patch on its head. The fish may be a *Tancho Showa*, *Tancho Sanke*, or even *Tancho Goshiki.* It is named for the Japanese crane (*Grus japonensis*), which also has a red spot on its head.
- *Chagoi*, "tea-coloured", this koi can range in colour from pale olive-drab green or brown to copper or bronze and more recently, darker, subdued orange shades. Famous for its docile, friendly personality and large size, it is considered a sign of good luck among koi keepers.
- *Asagi* koi is light blue above and usually red below, but also occasionally pale yellow or cream, generally below the lateral line and on the cheeks. The Japanese name means pale greenish-blue, spring onioncolour, or indigo. Sometimes it is incorrectly written as EmÄž (light yellow).
- *Utsurimono* is a black koi with a white, red, or yellow markings, in a zebra colour pattern. The oldest attested form is the yellow form, called "black and white markings" in the 19th century, but renamed *Ki Utsuri* by Elizaburo Hoshino, an early 20th-century koi breeder. The red and white

versions are called *Hi Utsuri* and *Shiro Utsuri* respectively. The word *utsuri* means to print (the black markings are reminiscent of ink stains). Genetically, it is the same as *Showa*, but lacking either red pigment (*Shiro Utsuri*) or white pigment (*Hi Utsuri/Ki Utsuri*).

- *Bekko* is a white-, red-, or yellow-skinned koi with black markings *sumi*. The Japanese name means "tortoise shell," and is commonly written as y0c02u. The white, red, and yellow varieties are called *Shiro Bekko*, *Aka Bekko* and *Ki Bekko*, respectively. It may be confused with the *Utsuri*.
- *Goshiki* is a dark koi with red (*Kohaku* style) *hi* pattern. The Japanese name means "five colours". It appears similar to an *Asagi*, with little or no *hi* below the lateral line and a *Kohaku Hi* pattern over reticulated (fishnet pattern) scales. The base colour can range from nearly black to very pale, sky blue.
- *Shûsui* means "autumn green"; the *Shûsui* was created in 1910 by Yoshigoro Akiyamaÿ, by crossing Japanese *Asagi* with German mirror carp. The fish has no scales, except for a single line of large mirror scales dorsally, extending from head to tail. The most common type of *Shûsui* have a pale, sky-blue/gray colour above the lateral line and red or orange (and very, very rarely bright yellow) below the lateral line and on the cheeks.
- *Kinginrin* is a koi with metallic (glittering, metal-flake-appearing) scales. The name translates into English as "gold and silver scales"; it is often abbreviated to *Ginrin*. There are *Ginrin* versions of almost all other varieties of koi, and they are fashionable. Their sparkling, glittering scales contast to the smooth, even, metallic skin and scales seen in the Ogon varieties. Recently, these characteristics have been combined to create the new *ginrin Ogon* varieties.
- *Kawarimono* is a "catch-all" term for koi that cannot be put into one of the other categories. This is a competition category, and many new varieties of koi compete in this one category. It is also known as *kawarigoi*.
- *Ôgon* is a metallic koi of one colour only. The most commonly encountered colours are gold, platinum, and orange. Cream specimens are very rare. *Ogon* compete in the *Kawarimono* category and the Japanese name means "gold." The variety was created by Sawata Aoki in 1946 from wild carp he caught in 1921. Recently, the metallic-skinned *Ogon* is being crossed with *ginrin*-scaled fish to create the *ginrin Ogon* with metallic skin and sparkling (metal flake) scales.
- *Kumonryû* literally "nine tattooed dragons" ÿ is a black *doitsu*-scaled fish with curling white markings. The patterns are thought to be reminiscent of Japanese ink paintings of dragons. They famously change colour with the seasons. *Kumonryu* compete in the *Kawarimono* category.

- *Ochiba* is a light blue/gray koi with copper, bronze, or yellow (*Kohaku*-style) pattern, reminiscent of autumn leaves on water. The Japanese name means "fallen leaves".
- *Koromo* is a white fish with a *Kohaku*-style pattern with blue or black-edged scales only over the *hi* pattern. This variety first arose in the 1950s as a cross between a *Kohaku* and an *Asagi*. The most commonly encountered *Koromo* is an *Ai Goromo*, which is coloured like a *Kohaku*, except each of the scales within the red patches has a blue or black edge to it. Less common is the *Budo-Goromo*, which has a darker (burgundy) *hi* overlay that gives it the appearance of bunches of grapes. Very rarely seen is the *Tsumi-Goromo* which is similar to *Budo-Goromo*, but the *hi* pattern is such a dark burgundy that it appears nearly black.
- *Hikari-moyomono* is a koi with coloured markings over a metallic base or in two metallic colours.
- *Kikokuryûÿ*", literally "sparkle" or "glitter black dragon" ÿis a metallic-skinned version of the *Kumonryu*.
- *Kin-Kikokuryû*, literally "gold sparkle black dragon" or "gold glitter black dragon" is a metallic-skinned version of the *Kumonryu* with a *Kohaku*-style *hi* pattern developed by Mr. Seiki Igarashi of Ojiya City. There are (at least) six different genetic subvarieties of this general variety.
- Ghost ko, a hybrid of *Ogon* and wild carp with metallic scales, is considered by some to be not *nishikigoi*.
- Butterfly is a hybrid of koi and Asian carp with long flowing fins. Various colourations depend on the koi stock used to cross. It also is considered by some to not be *nishikigoi*.
- *Doitsu-goi* originated by crossbreeding numerous different established varieties with "scaleless" German carp (generally, fish with only a single line of scales along each side of the dorsal fin). Also written as ìr8 É›, there are four main types of *Doitsu* scale patterns. The most common type (referred to above) has a row of scales beginning at the front of the dorsal fin and ending at the end of the dorsal fin (along both sides of the fin). The second type has a row of scales beginning where the head meets the shoulder and running the entire length of the fish (along both sides). The third type is the same as the second, with the addition of a line of (often quite large) scales running along the lateral line (along the side) of the fish, also referred to as "mirror koi". The fourth (and rarest) type is referred to as "armour koi" and are completely (or nearly) covered with very large scales that resemble plates of armour.

Differences from Goldfish

Goldfish were developed in China more than a thousand years ago by selectively breeding Prussian carp for colour mutations. By the Song Dynasty (960 – 1279), yellow, orange, white and red-and-white colourations had been developed. Goldfish (*Carassius auratus*) and Prussian carp (*Carassius gibelio*) are now considered different species. Goldfish were introduced to Japan in the 16th century and to Europe in the 17th century. Koi, on the other hand, were developed from common carp in Japan in the 1820s. Koi are domesticated common carp (*Cyprinus carpio*) that are selected or culled for colour; they are not a different species, and will revert to the original colouration within a few generations if allowed to breed freely.

In general, goldfish tend to be smaller than koi, and have a greater variety of body shapes, and fin and tail configurations. Koi varieties tend to have a common body shape, but have a greater variety of colouration and colour patterns. They also have prominent barbels on the lip. Some goldfish varieties, such as the common goldfish, comet goldfish and shubunkin have body shapes and colouration that are similar to koi, and can be difficult to tell apart from koi when immature. Since goldfish and koi were developed from different species of carp, even though they can interbreed, their offspring are sterile.

Health, Maintenance and Longevity

The common carp is a hardy fish, and koi retain that durability. Koi are cold-water fish, but benefit from being kept in the 15-25 °C (59-77°F) range, and do not react well to long, cold, winter temperatures; their immune systems "turn off" below 10°C. Koi ponds usually have a metre or more of depth in areas of the world that become warm during the summer, whereas in areas that have harsher winters, ponds generally have a minimum of 1.5 metres (4½ feet). Specific pond construction has evolved by koi keepers intent on raising show-quality koi.

Koi's bright colours put them at a severe disadvantage against predators; a white-skinned *Kohaku* is a visual dinner bell against the dark green of a pond. Herons, kingfishers, otters, raccoons, cats, foxes, badgers and hedgehogs are all capable of emptying a pond of its fish. A well-designed outdoor pond will have areas too deep for herons to stand, overhangs high enough above the water that mammals cannot reach in, and shade trees overhead to block the view of aerial passers-by. It may prove necessary to string nets or wires above the surface. A pond usually includes a pump and filtration system to keep the water clear.

Koi are an omnivorous fish, and will eat a wide variety of foods, including peas, lettuce, and watermelon. Koi food is designed not only to be nutritionally balanced, but also to float so as to encourage them to come to the surface. When they are eating, it is possible to check koi for parasites and ulcers. Koi will recognise the persons feeding them and gather around them at feeding times. They can be trained to take food from one's hand. In the winter, their digestive systems slow nearly to a halt, and they eat very little, perhaps no more than nibbles of algae from the bottom. Care should be taken by hobbyists that proper oxygenation and off-gassing occurs over the winter months in small water ponds, so they do not perish. Their appetites will not come back until the water becomes warm in the spring. When the temperature drops below 50°F (10°C), feeding, particularly with protein, is halted or the food can spoil in their stomachs, causing sickness.

One famous scarlet koi, named "Hanako" (c. 1751 – July 7, 1977) was owned by several individuals, the last of whom was Dr. Komei Koshihara. Hanako was supposedly 226 years old upon her death, based on examining one of her scales in 1966. Koi "maximum longevity" is listed as 47 years old.

Breeding

Like most fish, koi reproduce through spawning in which a female lays a vast number of eggs and one or more males fertilize them. Nurturing the resulting offspring (referred to as "fry") is a tricky and tedious job, usually done only by professionals. Although a koi breeder may carefully select the parents they wish based on their desired characteristics, the resulting fry will nonetheless exhibit a wide range of colour and quality.

Koi will produce thousands of offspring from a single spawning. However, unlike cattle, purebred dogs, or more relevantly, goldfish, the large majority of these offspring, even from the best champion-grade koi, will not be acceptable as *nishikigoi* (they have no interesting colours) or may even be genetically defective. These unacceptable offspring are culled at various stages of development based on the breeder's expert eye and closely guarded trade techniques. Culled fry are usually destroyed or used as feeder fish (mostly used for feeding arowana due to the belief it will enhance its colour), while older culls, within their first year between 3" to 6" long (also called "Tosai"), are often sold as lower-grade, pond-quality koi.

The semirandomised result of the koi's reproductive process has both advantages and disadvantages for the breeder. While it requires diligent oversight to narrow down the favourable result the breeder wants, it also makes possible the development of new varieties of koi within relatively few generations.

In the Wild

Koi have been accidentally or deliberately released into the wild in every continent except Antarctica. They quickly revert to the natural colouration of common carp within a few generations. In many areas, they are considered an invasive species and pests. They greatly increase the turbidity of the water because they are constantly stirring up the substrate. This makes waterways unattractive, reduces the abundance of aquatic plants, and can render the water unsuitable for swimming or drinking, even by livestock. In some countries, koi have caused so much damage to waterways that vast amounts of money and effort have been spent trying to eradicate them, largely unsuccessfully.

Setu Babakan

Setu Babakan or Small Lake Babakan (Setu or Situ means Small Lake) is located in Srengseng Sawah, Sub-district of Jagakarsa, Municipality of South Jakarta, Indonesia near Depok or around 5 kilometres southeast of Ragunan Zoo at the centre of the Betawi Cultural Village, a site considered part of the cultural heritage of Jakarta, which is devoted to the preservation of the indigenous Betawi culture. The location of the Betawi Cultural Village is replacement of the previous Condet (Betawi) Cultural Village which is eroded by the time.

The Situ Babakan Lake has an area of 32 hectares (79 ac) in which the water flows in from the Ciliwung River and currently is used for fish farming by the Betawi people who live in the vicinity of the lake.

There are more than 100 floating net cages placed in the lake to breed different kinds of fish, including carp, tilapia, and several species of ornamental fish. The lake is a place for aquatic recreational activities, such as rafting, and fishing. The garden surrounding the lake is cultivated with fruit plants, such as banana, coconut, and guava.

Hatchery

A hatchery is a facility where eggs are hatched under artificial conditions, especially those of fish or poultry. It may be used for ex-situ conservation purposes, i.e. to breedrare or endangered species under controlled conditions; alternatively, it may be for economic reasons (i.e. to enhance food supplies or fishery resources).

Fish hatcheries

Fish hatcheries are used to cultivate and breed a large number of fish in an enclosed environment. Fish farms use hatcheries to cultivate fish to sell for food, or ornamental purposes, eliminating the need to find the fish in the wild and even providing some species outside of their natural season. They raise the fish until they are ready to be eaten or sold to aquarium stores. Other hatcheries release the juvenile fish into a river, lake or the ocean to support commercial, tribal, or recreational fishing or to supplement the natural numbers of threatened or endangered species, a practice known as fish stocking.

Researchers have raised concerns about hatchery fish potentially breeding with wild fish. Hatchery fish may in some cases compete with wild fish. In the United States and Canada, there have been several salmon and steelhead hatchery reform projects intended to reduce the possibility of negative impacts from hatchery programs. Most salmon and steelhead hatcheries are managed better and follow up to date management practices to ensure any risks are minimised.

Poultry Hatcheries

Poultry hatcheries produce a majority of the birds consumed in the developed world including chickens, turkeys, ducks and some other minor bird species. It is a multibillion dollar industry, with highly regimented production systems used to maximise bird size versus feed consumed. Birds are produced and maintained under high density, which makes production and harvesting more economical, but can also generate problems such as the spread of pathogens, which can move very quickly through the population when animal densities are high.

Poultry generally start with naturally (chickens) or artificially (turkeys) inseminated hens that lay eggs; the eggs are cleaned and shells are checked for soundness before putting them in the incubators. The incubators control temperature and humidity, and turn the eggs until they hatch. Generally large numbers are produced at one time so the resulting birds are uniform in size and can be harvested at the same time. Once the eggs hatch and the chicks are a few days old, they are often vaccinated, beak-trimmed and or toe-clipped; this involves the removal of half of the top beak and the clipping of the toe ends. This is done to prevent the birds from harming each other while they are living in close proximity to each other. After these procedures, they are moved to enclosed buildings to be raised until harvest.

For chickens bred as (egg) layers, only the female chicks are considered to be of value; in excess of 100,000 male chicks can be dropped into an industrial "grinder" and disposed of in a single day.

Shrimp

Shrimp are swimming, decapod crustaceans classified in the infraorder Caridea, found widely around the world in both fresh and salt water. Adult shrimp are filter feeding benthic animals living close to the bottom. They can live in schools and can swim rapidly backwards. Shrimp are an important food source for larger animals from fish to whales. They have a high tolerance to toxins in polluted areas, and may contribute to high toxin levels in their predators. Together with prawns, shrimp are widely caught and farmed for human consumption.

Taxonomy

A number of more or less unrelated crustaceans share the word "shrimp" in their common name. Examples are the mantis shrimp and the opossum or mysid shrimp, both of which belong to the same class (Malacostraca) as decapods including shrimp, but constitute two different orders within it, the Stomatopoda and the Mysidacea. *Triops longicaudatus* and *Triops cancriformis* are also popular animals in freshwater aquaria, and are often called shrimp, although they belong instead to the Notostraca, a quite unrelated group.

Biological Definition

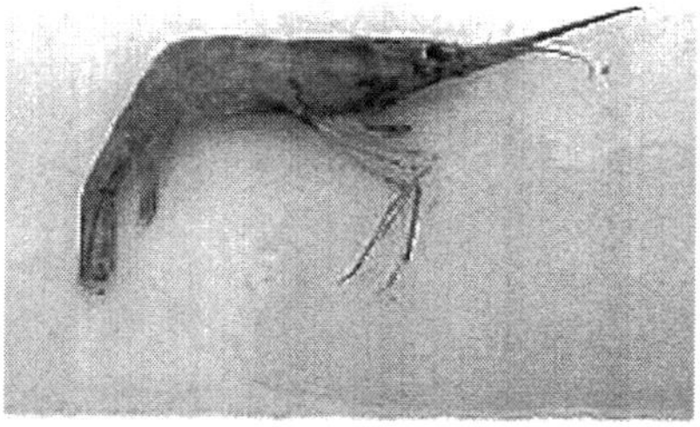

Figure: Shrimp (Caridea), such as *Pandalus borealis*, typically have two pairs of claws, and the second segment of the abdomen overlaps the segments on either side. The abdomen shows a pronounced *caridean bend.*

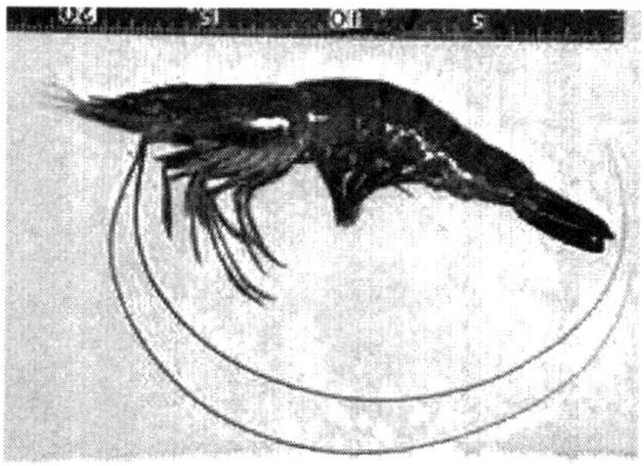

Figure: Prawns (Dendrobanchiata), such as *Penaeus monodon*, typically have three pairs of claws, and even-sized segments on the abdomen. There is no pronounced bend in the abdomen.

The class Malacostraca contains about half of the crustaceans. The members of this class have a primitive body plan that can be described as shrimp-like, consisting of a 5-8-7 body plan. They have a small carapace that encloses the head and the thorax, and have a muscular abdomen for swimming. They also have a thin exoskeleton to maintain a light weight. These general characteristics are common in all members of the class.

The class can be further divided into the decapods, which are even still divided into the dendrobranchiates (prawns) and the carideans (shrimp and snapping shrimp).

The prawns have sequentially overlapping body segments (segment one covers the segment two, segment two covers segment three, etc.), chelate (claw like) first three leg pairs, and have a very basic larval body type.

The shrimps also have overlapping segments, however, in a different pattern (segment two overlaps segments one and three), only the first two leg pairs are chelate, and they have a more complex larval form.

Biologists distinguish the true shrimp from the true prawn because of the differences in their gill structures. The gill structure is lamellar in shrimp but branching in prawns. The easiest practical way to separate true shrimps from true prawns is to examine the second abdominal segment. The second segment of a shrimp overlaps both the first and the third segment, while the second segment of a prawn overlaps only the third segment.

The infraorder Caridea is divided into 16 superfamilies:

- Alpheoidea Rafinesque, 1815
- Atyoidea De Haan, 1849
- Bresilioidea Calman, 1896
- Campylonotoidea Sollaud, 1913
- Crangonoidea Haworth, 1825
- Galatheacaridoidea Vereshchaka, 1997
- Nematocarcinoidea Smith, 1884
- Oplophoroidea Dana, 1852
- Palaemonoidea Rafinesque, 1815
- Pandaloidea Haworth, 1825
- Pasiphaeoidea Dana, 1852
- Physetocaridoidea Chace, 1940a
- Procaridoidea Chace & Manning, 1972
- Processoidea Ortmann, 1896
- Psalidopodoidea Wood-Mason, 1892
- Stylodactyloidea Bate, 1888

A number of extinct genera cannot be placed in any superfamily:

- *Acanthinopus* Pinna, 1974
- *Alcmonacaris* Polz, 2009
- *Bannikovia* Garassino & Teruzzi, 1996
- *Blaculla* Münster, 1839
- *Buergerocaris* Schweigert & Garassino, 2004
- *Gampsurus* von der Marck, 1863
- *Hefriga* Münster, 1839

- *Leiothorax* Pinna, 1974
- *Parvocaris* Bravi & Garassino, 1998
- *Pinnacaris* Garassino & Teruzzi, 1993

Commercial and Culinary Definition

While in biological terms shrimps and prawns belong to different suborders of Decapoda, they are very similar in appearance. In commercial farming and fisheries, the terms "shrimp" and "prawn" are often used interchangeably. However, recent aquaculture literature increasingly uses the term "prawn" only for the freshwater forms of palaemonids and "shrimp" for the marine penaeids.

In the United Kingdom, the word "prawn" is more common on menus than "shrimp"; while the opposite is the case in North America. The term "prawn" is also loosely used to describe any large shrimp, especially those that come 15 (or fewer) to the pound (such as "king prawns", yet sometimes known as "jumbo shrimp"). Australia and some other Commonwealth nations follow this British usage to an even greater extent, using the word "prawn" almost exclusively. When Australian comedian Paul Hogan used the phrase, "I'll slip an extra shrimp on the barbie for you" in an American television advertisement, it was intended to make what he was saying easier for his American audience to understand, and was thus a deliberate distortion of what an Australian would typically say.

In Britain very small crustaceans with a brownish shell are called shrimp, and are used to make potted shrimp. They are also used in dishes where they are not the primary ingredient.

Consumption

As with other seafood, shrimp is high in calcium, iodine and protein but low in food energy. A shrimp-based meal is also a significant source of cholesterol, from 122 mg to 251 mg per 100 g of shrimp, depending on the method of preparation. Shrimp consumption, however, is considered healthy for the circulatory system because the lack of significant levels of saturated fat in shrimp means that the high cholesterol content in shrimp actually improves the ratio of LDL to HDL cholesterol and lowers triglycerides. Shrimp and other shellfish are among the most common food allergens. They are not kosher and thus are forbidden in Jewish cuisine. On the other hand, shrimp are halalaccording to some madhâhib, and therefore are permissible in Islamic cuisine.

Commercial Fishing

Common commercial methods for catching shrimp and prawns include otter trawls, cast nets, seines, shrimp baiting and dip netting. Trawling involves the use of a system of nets. In some parts of the Pacific Northwest, fishing with baited traps is also common.

The following table shows the yearly weight of shrimp and prawns captured globally in millions of tonnes:

Production	1999	2000	2001	2002	2003	2004	2005
Million Tonnes	3.03	3.09	2.96	2.97	3.55	3.54	3.42

The highest rates of incidental catch of non-target species is associated with shrimp trawling. In 1997, the FAO documented the estimated by catch and discard levels from shrimp fisheries around the world. They found discard rates as high as 20 pounds for every pound of shrimp, with a world average of 5.7 pounds for every pound of shrimp.

Trawl nets in general, and shrimp trawls in particular, have been identified as sources of mortality for species of finfish and cetaceans. Bycatch is often discarded dead or dying by the time it is returned to the sea, and may alter the ecological balance in discarded regions. Worldwide, shrimp trawl fisheries generate about 2% of the world's catch of fish in weight, but result in more than one third of the global bycatch total.

Farming

A shrimp farm is an aquaculture business for the cultivation of marine shrimp or prawns for human consumption. Commercial shrimp farming began in the 1970s, and production grew steeply, particularly to match the market demands of the United States, Japan and Western Europe. The total global production of farmed shrimp reached more than 1.6 million tonnes in 2003, representing a value of nearly 9 billion U.S. dollars. About 75% of farmed shrimp are produced in Asia, in particular in China, Thailand and in the Philippines (normally shrimp or prawns are caught in lake, river and sea and are not farmed in tanks). The other 25% are produced mainly in Latin America, where Brazil is the largest producer. The largest exporting nation is Thailand. In Southeast Asia, mangroves are cotinually taken over then discarded after several years, leading to the equivalent of a ton of carbon dioxide emissions per pound of shrimp produced.

Marketing

Shrimp are marketed and commercialised with several issues in mind. Most shrimp are sold frozen and marketed based on their categorisation of presentation, grading, colour, and uniformity.

Preparation

Preparing shrimp for consumption usually involves removing the head, shell, tail, and "sand vein". To de-shell a shrimp, the tail is held while gently removing the shell around the body. The tail can be detached completely at this point, or left attached for presentation purposes. Removing the "sand vein" (a euphemism for the digestive tract) is referred to as "deveining". The sand vein can be removed by making a shallow cut lengthwise down the outer curve of the shrimp's body, allowing the dark ribbon-like digestive tract to be removed with a pointed utensil. Alternatively, if the tail has been detached, the vein can be pinched at the tail end and pulled out completely with the fingers. On large shrimp, the "blood vein" (a euphemism for the ventral nerve cord) along the inner curve of the shrimp's body is typically removed as well. The shrimp is then rinsed under cold running water.

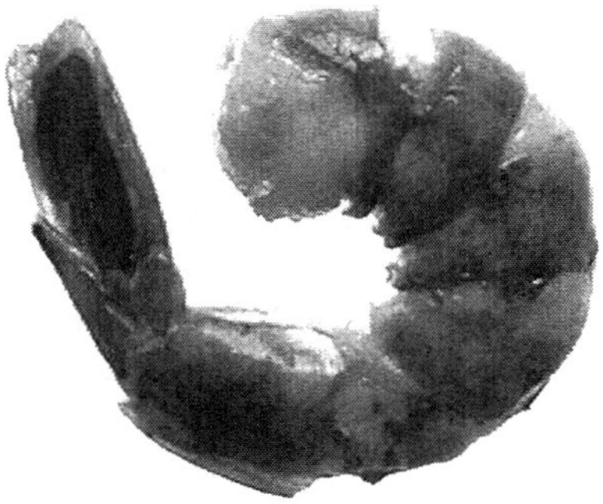

Figure: A steamed tail-on shrimp

Shrimp and prawns are versatile ingredients, and are often used as an accompaniment to fried rice. Common methods of preparation include baking, boiling, frying, and grilling.

Recipes using shrimp form part of the cuisine of many cultures. Strictly speaking, dishes containing scampi should be made from the Norway lobster, a shrimp-like crustacean more closely related to the lobster than shrimp, but in some places it is quite common for large shrimp to be used instead.

Wet shrimp is commonly used as a flavouring and as a soup base in Asian cuisines while fried shrimp is popular in North America. In Europe, shrimp is very popular, forming a necessary ingredient in Spanish *paella de marisco*, Italian *cacciucco*, Portuguese *caldeirada* and many other seafood dishes. Shrimp curry is very popular in South Asia and Southeast Asia. Shrimp are also found in Latin and Caribbean dishes such as enchiladas and coconut shrimp. Other recipes include jambalaya, okonomiyaki, poon choi andbagoong. Shrimp are also consumed as salad, by frying, with rice, and as shrimp guvec (a dish baked in a clay pot) in the Western and Southern coasts of Turkey.

Deveining

Deveining is the removal of the gastrointestinal tract of a shrimp, a common part of preparing them for human consumption. The digestive tract is a dark band running from the head to the tail of the animal, where the spine would be if they were vertebrates. In females, the reproductive canal is also in the same area.

One must first cut a slit in the shell at the back of the shrimp. Special deveining tools are sometimes used but knives, skewers, and even toothpicks can be used to devein.

Removing the vein is not essential, as it is not poisonous and is mostly tasteless. Deveining does slightly change the flavour and makes it more consistent. Shrimp also sometimes consume small amounts of sand by accident and the vein thus might be gritty.

Life Cycle

Most shrimp mature and breed only in a marine habitat, although there are a small number of freshwater species. The females lay 50,000 to 1 million eggs, which hatch after some 24 hours into tiny nauplii. These nauplii feed on yolk reserves within their body and then undergo a metamorphosis into zoeae. This second larval stage feeds in the wild on algae and after a few days metamorphoses again into the third stage to become myses.

At this stage the myses already begin to appear like tiny versions of fully developed adults and feed on algae and zooplankton. After another three to four days they metamorphose a final time into postlarvae: young shrimp having all the characteristics of adults. The whole process takes about 12 days from hatching. In the wild, the marine postlarvae then migrate into estuaries, which are rich in nutrients and low in salinity. There they grow and eventually migrate back into open waters when they mature. Most adult shrimp are benthic animals living primarily on the sea floor.

Common shrimp species include pink, brown and snapping shrimp. Depending on the species and location, they grow from about 1.2 to 30 centimetres (0.47 to 12 in) long, and live between one and 6.5 years.

Fossil Record

The fossil record of shrimp is sparse, with only 57 exclusively fossil species known. The earliest of these cannot be assigned to any family, but date from the Lower Jurassic and Cretaceous.

Home Aquaria

Several types of shrimp are kept in home aquaria. Some are purely ornamental, while others are useful in controlling algae and removing debris. Freshwater shrimp commonly available for aquaria include the Bamboo shrimp, Japanese marsh shrimp (*Caridina multidentata,* also called "Amano shrimp," as their use in aquaria was pioneered by Takashi Amano), cherry shrimp (*Neocaridina heteropoda*), and ghost or glass shrimp (*Palaemonetes* spp.). Popular saltwater shrimp include the cleaner shrimp *Lysmata amboinensis*, the fire shrimp (*Lysmata debelius*) and the harlequin shrimp (*Hymenocera picta*).

Etymology

The term *shrimp* originated around the 14th century with the Middle English *shrimpe*, akin to the Middle Low German *schrempen*, and meaning to contract or wrinkle; and the Old Norse *skorpna*, meaning to shrivel up.

Cichlids are fishes from the family Cichlidae in the order Perciformes. Cichlids are members of a group known as the Labroidei along with the wrasses (Labridae), damselfish (Pomacentridae), and surfperches (Embiotocidae). This family is both large and diverse. At least 1,650 species have been scientifically described, making it one of the largest vertebrate families. New species are discovered annually, and many species remain undescribed. The actual number of species is therefore unknown, with estimates varying between 1,300 and 3,000.

Description

Cichlids span a wide range of body sizes, from species as small as 2.5 centimetres (0.98 in) in length (e.g., female *Neolamprologus multifasciatus*) to much larger species approaching 1 metre (3.3 ft) in length (e.g. *Boulengerochromis* and *Cichla*). As a group, cichlids exhibit a similar diversity of body shapes, ranging from strongly laterally compressed species (such as *Altolamprologus*, *Pterophyllum*, and *Symphysodon*) to species that are cylindrical and highly elongate (such as *Julidochromis*, *Teleogramma*,*Teleocichla*, *Crenicichla*, and *Gobiocichla*). Generally, however, cichlids tend to be of medium size, ovate in shape and slightly laterally compressed, and generally similar to the North American sunfishes in morphology, behaviour, and ecology.

Many cichlids, particularly tilapia, are important food fishes, while others are valued game fish (e.g. *Cichla* species). The family also includes many familiar aquarium fish, including the angelfish, oscars, and discus. Cichlids have the largest number of endangered species among vertebrate families, most in the haplochromine group.Cichlids are particularly well known for having evolved rapidly into a large number of closely related but morphologically diverse species within large lakes, particularlyTanganyika, Victoria, Malawi, and Edward. Their diversity in the African Great Lakes is important for the study of speciation in evolution. Many cichlids that have been introduced into waters outside of their natural range have become nuisances, such as tilapia in the southern United States.

Anatomy and Appearance

Cichlids share a single key trait: the fusion of the lower pharyngeal bones into a single tooth-bearing structure. A complex set of muscles allows the upper and lower pharyngeal bones to be used as a second set of jaws for processing food, allowing a division of labour between the "true jaws" (mandibles) and the "pharyngeal jaws". Cichlids are efficient feeders that capture and process a very wide variety of food items. This is assumed to be one reason why they are so diverse. Cichlids vary in body shape, ranging from compressed and disc-shaped (such as *Symphysodon*), to triangular (such as *Pterophyllum)*, to elongate and cylindrical (such as *Crenicichla*).

The features that distinguish them from the other Labroidei include:

- A single nostril on each side of the forehead, instead of two.
- No bony shelf below the orbit of the eye.
- Division of the lateral line organ into two sections, one on the upper half of the flank and a second along the midline of the flank from about halfway along the body to the base of the tail (except for genera *Teleogramma* and *Gobiocichla*).
- A distinctively shaped otolith.
- The small intestine's left-side exit from the stomach instead of its right side as in other Labroidei.

Taxonomy

Kullander (1998) recognises eight subfamilies of cichlids: the *Astronotinae, Cichlasomatinae, Cichlinae, Etroplinae, Geophaginae, Heterochromidinae, Pseudocrenilabrinae*, and *Retroculinae.* A ninth subfamily, Ptychochrominae, was later recognised by Sparks and Smith. Cichlid taxonomy is still debated, and classification of genera cannot yet be definitively given. A comprehensive system of assigning species to monophyletic genera is still lacking, and there is not complete agreement on what genera should be recognised in this family.

As an example of the classification problems, Kullander placed the African genus *Heterochromis* phylogenetically within neotropical cichlids, although later papers concluded otherwise. Other problems centre upon the identity of the putative common ancestor for the Lake Victoria superflock, and the ancestral lineages of Tanganyikan cichlids.

Comparisons between a morphologically-based phylogeny and analyses of gene loci produce differences at the genus level. There remains a consensus that the *Cichlidae* as a family is monophyletic.

One problem that transformed cichlid taxonomy is related to dentition, which had been used as a classifying characteristic. In many cichlids, tooth shape changes with age, due to wear, and cannot be relied upon. Genome sequencing and other technologies transformed cichlid taxonomy.

Range and Habitat

Cichlids are the most species-rich non-Ostariophysan family in freshwaters worldwide. They are most diverse in Africa and South America. It is estimated that Africa alone hosts at least 1,600 species. Central America and Mexico have approximately 120 species, as far north as the Rio Grande in southern Texas. Madagascar has its own distinctive species (*Oxylapia, Paratilapia, Paretroplus, Ptychochromis,* and *Ptychochromoides*), only distantly related to those on the African mainland. Native cichlids are largely absent in Asia, except for nine species in Israel, Lebanon and Syria (*Astatotilapia flaviijosephi, Oreochromis aureus, O. niloticus, Sarotherodon galilaeus, Tilapia zillii,* and *Tristramella spp.*), one in Iran (*Iranocichla*), and three in India and Sri Lanka (*Etroplus*). If disregarding Trinidad and Tobago (where the few native cichlids are members of genera that are widespread in the South American mainland), the three species from the genus *Nandopsis* are the only cichlids from the Antilles in the Caribbean, specifically Cuba and Hispaniola. Europe, Australia, Antarctica, and North America north of the Rio Grande drainage have no native cichlids, although in Florida, Mexico, Japan and northern Australia feral populations of cichlids have become established as exotics.

Although most cichlids are found at relatively shallow depths, several exceptions do exist. These include species such as *Alticorpus macrocleithrum* and *Pallidochromis tokolosh* down to 150 metres (490 ft) below the surface in Lake Malawi, and the whitish (non-pigmented) and blind *Lamprologus lethops*, which is believed to live as deep as 160 metres (520 ft) below the surface in the Congo River.

Cichlids are less commonly found in brackish and saltwater habitats, though many species tolerate brackish water for extended periods; *Cichlasoma urophthalmus*, for example, is equally at home in freshwater marshes and mangrove swamps, and lives and breeds in saltwater environments such as the mangrove belts around barrier islands. Several species of *Tilapia, Sarotherodon,* and *Oreochromis* are euryhaline and can disperse along brackish coastlines between rivers. Only a few cichlids, however, inhabit primarily brackish or salt water, most notably *Etroplus maculatus, Etroplus suratensis,* and *Sarotherodon melanotheron*. The perhaps most extreme habitats for cichlids are the warm hypersaline lakes where the members of the genera *Alcolapia* and *Danakilia*are found. Lake Abaeded in Eritrea encompasses the entire distribution of *D. dinicolai*, and its temperature ranges from 29 to 45 °C (84 to 113 °F).

With the exception of the species from Cuba and Hispaniola, cichlids have not reached any oceanic island and have a predominantly Gondwanan distribution, showing the precise sister relationships predicted by vicariance: Africa-South America and India-Madagascar. The dispersal hypothesis, in contrast, requires cichlids to have negotiated thousands of kilometres of open ocean between India and Madagascar without colonising any other island or, for that matter, crossing the Mozambique Channel to Africa. Although the vast majority of Malagasy cichlids are entirely restricted to freshwater, *Ptychochromis grandidieri* and

Paretroplus polyactis are commonly found in coastal brackish water and they are apparently salt tolerant, as is also the case for *Etroplus maculatus* and *E. suratensis* from India and Sri Lanka.

Ecology

Feeding

Many cichlids are primarily herbivores feeding on algae (e.g. *Petrochromis*) and plants (e.g. *Etroplus suratensis*). Small animals, particularly invertebrates, are only a minor part of their diet.

Other cichlids are detritivores and eat all types of organic material; called Aufwuchs; among these species are the tilapiines of the genera *Oreochromis*, *Sarotherodon*, and *Tilapia*.

Other cichlids are predatory and eat little or no plant matter. These include generalists that catch a variety of small animals, including other fishes and insect larvae (e.g.*Pterophyllum*), as well as variety of specialists. *Trematocranus* is a specialised snail-eater, while *Pungu maclareni* feeds on sponges. A number of cichlids feed on other fish, either entirely or in part. *Crenicichla* are stealth-predators that lunge from concealment at passing small fish, while *Rhamphochromis* are open water pursuit predators that chase down their prey. Paedophagous cichlids such as the *Caprichromis* species eat other species' eggs or young, in some cases ramming the heads of mouth brooding species to force them to disgorge their young. Among the more unusual feeding strategies are those of *Corematodus*, *Docimodus evelynae*, *Plecodus*, *Perissodus*, and *Genyochromis*spp., which feed on scales and fins of other fishes, a behaviour known as lepidophagy, along with the death-mimicking behaviour of *Nimbochromis* and *Parachromis*species, which lay motionless, luring small fish to their side prior to ambush.

This variety of feeding styles has helped cichlids to inhabit similarly varied habitats. Its pharyngeal teeth (teeth in the throat) afford cichlids so many "niche" feeding strategies, because the jaws pick and hold food, while the pharyngeal teeth crush the prey.

Reproduction

Brood Care

All species show some form of parental care for both eggs and larvae, often nurturing free-swimming young until they are weeks or months old.

Communal parental care, where multiple monogamous pairs care for a mixed school of young have also been observed in multiple cichlid species, including *Amphilophus citrinellus*, *Etroplus suratensis*, and *Tilapia rendalli*. Comparably, the fry of *Neolamprologus brichardi*, a species that commonly lives in large groups, are protected not only by the adults, but also by older juveniles from previous spawns.

Several cichlids, including discus (*Symphysodon* spp.), some *Amphilophus* species, *Etroplus* and *Uaru* species feed their young with a skin secretion from mucous glands.

Parental care falls into one of four categories: substrate or open brooders, secretive cave brooders (also known as guarding speleophils), and at least two types of mouthbrooders, ovophile mouthbrooders and larvophile mouthbrooders.

Open Brooding

Open or substrate brooding cichlids lay their eggs in the open, on rocks, leaves, or logs. Examples of open brooding cichlids include *Pterophyllum*, *Symphysodon* spp, and Anomalochromis *thomasi*. Male and female parents usually engage in differing brooding roles. Most commonly, the male patrols the pair's territory and repels intruders, while females fan water over the eggs, removing the infertile and leading the fry while foraging. However, both sexes are able to perform the full range of parenting behaviours.

Cave Brooding

A female *Cyphotilapia frontosa* mouthbrooding fry, which can be seen looking out her mouth

Secretive cave spawning cichlids lay their eggs in caves, crevices, holes, or discarded mollusc shells, frequently attaching the eggs to the roof of the chamber. Examples include *Pelvicachromis* spp. *Archocentrus* spp, and *Apistogramma* spp.Free-swimming fry and parents communicate in captivity and in the wild. Frequently this communication is based on body movements, such as shaking and pelvic fin flicking. In addition, open and cave brooding parents assist in finding food resources for their fry. Multiple neotropical cichlid species perform leaf-turning and fin-digging behaviours.

Ovophile Mouthbrooding

Ovophile mouthbrooders incubate their eggs in their mouths as soon as they are laid, and frequently mouthbrood free-swimming fry for several weeks. Examples include many East African Rift lakes (Lake Malawi, Lake Tanganyika and Lake Victoria) endemics, e.g.: *Maylandia*, *Pseudotropheus*, and *Tropheus*, along with some South American cichlids such as *Geophagus steindachneri*.

Larvophile Mouthbrooding

Larvophile mouthbrooders lay eggs in the open or in a cave and take the hatched larvae into the mouth. Examples include some variants of *Geophagus altifrons*, and some *Aequidens*, *Gymnogeophagus*, and *Satanoperca*. Mouthbrooders, whether of eggs or larvae, are predominantly females. Exceptions that also involve the males include eretmodine cichlids (genera *Spathodus*, *Eretmodus*, and *Tanganicodus*), some *Sarotherodon* species, *Chromidotilapia guentheri*, and some *Aequidens* species. Rare paternal mouthbrooding occurs, for example, in *Sarotherodon melanotheron*. This method appears to have evolved independently in several groups of African cichlids.

Mating

Cichlids mate either monogamously or polygamously. The mating system of a given cichlid species is not consistently associated with its brooding system. For example, although most monogamous cichlids are not mouthbrooders, *Chromidotilapia*, *Gymnogeophagus*, *Spathodus* and *Tanganicodus* are all monogamous mouthbrooders. In contrast, numerous open or cave spawning cichlids are polygamous; examples include *Apistogramma*, *Lamprologus*, *Nannacara* and *Pelvicachromis*.

Population Status

In 2010, the International Union for Conservation of Nature classified 184 species as vulnerable, 52 as endangered, and 106 as critically endangered. At present, the IUCN only lists *Yssichromis sp. nov. "argens"* asextinct in the wild, and six species are listed as entirely extinct, but it is acknowledged that many more possibly belong in these categories (for example, *Haplochromis aelocephalus*, *H. apogonoides*, *H. dentex*, *H. dichrourus* and numerous other members of the genus *Haplochromis* have not been seen since the 1980s, but are maintained as Critically Endangered in the small chance that tiny –but currently unknown– populations survive).

Lake Victoria

Because of the introduced Nile perch (*Lates niloticus*) and water hyacinth, deforestation that led to water siltation, and overfishing, many Lake Victoria species have been wiped out or drastically reduced. By around 1980, lake fisheries yielded only 1 percent cichlids, a drastic decline from 80 percent in earlier years.

About two-thirds of endemic cichlids (approximately 300 species), especially bottom feeders, became endangered or extinct. Some survivors have adapted by becoming smaller or hybridising with other species. Satellite lakes such as Lake Edward and Lake Kyoga have not been as strongly affected, however, and harbour an array of similar species.

Food and Game Fish

Although cichlids are mostly small- to medium-sized, many are notable as food and game fishes. With few thick rib bones and tasty flesh, artisan fishing is not uncommon in Central America and South America, as well as areas surrounding the African rift lakes.

Tilapia

The most important food cichlids, however, are the tilapiines of North Africa. Fast growing, tolerant of stocking density, and adaptable, tilapiine species have been introduced and farmed extensively in many parts of Asia and are increasingly common aquaculture targets elsewhere.

Farmed tilapia production is about 1,500,000 tonnes (1,500,000 long tons; 1,700,000 short tons) annually with an estimated value of US$1.8 billion, about equal to that of salmon and trout.

Unlike those carnivorous fish, tilapia can feed on algae or any plant-based food. This reduces the cost of tilapia farming, reduces fishing pressure on prey species, avoids concentrating toxins that accumulate at higher levels of the food chain and makes tilapia the preferred "aquatic chickens" of the trade.

Game Fish

Many large cichlids make good game fish. The strong, hard-fighting peacock bass (*Cichla* species) of South America is one of the most popular sportfish. It was introduced in many waters around the world. In Florida, this fish generates millions of hours of fishing and sportfishing revenue of more than US$8 million a year. Other cichlids preferred by anglers include the Oscar, Mayan cichlid (*Cichlasoma urophthalmus*), and jaguar guapote (*Parachromis managuensis*).

Aquarium Fish

Since 1945, cichlids have become increasingly popular as aquarium fish. Many cichlids are small to medium-sized, easy to feed with a range of prepared fish foods, breed readily, and practice brood care, making good aquarium fish. The most common species in hobbyist aquaria is *Pterophyllum scalare* from the Amazon River basin in tropical South America, known in the trade as the "angelfish".

Other popular or readily available species include the oscar (*Astronotus ocellatus*), convict cichlid (*Archocentrus nigrofasciatus*) and discus (*Symphysodon* spp.).

Cichlids can be kept in aquaria with other fish; however, many cichlids eat smaller fish. Conversely, some cichlids, such as *Apistogramma* or *Julidochromis* spp., can be timid. In such cases the use of dither fish is recommended.

Hybrids and Selective Breeding

Some cichlids readily hybridise with related species, both in the wild and under artificial conditions. Other groups of fishes, such as European cyprinids, also hybridise. Unusually, cichlid hybrids have been put to extensive commercial use, in particular for aquaculture and aquaria. The hybrid red strain of tilapia, for example, is often preferred in aquaculture for its rapid growth. Tilapia hybridisation can produce all-male populations to control stock density or prevent reproduction in ponds.

Aquarium Hybrids

The most ubiquitous aquarium hybrid is perhaps the blood parrot cichlid which is a cross of several species, especially from genus *Amphilophus*. With a beak-shaped mouth, an abnormal spine, and an occasionally missing caudal fin (known as the "love heart" parrot cichlid), the fish is controversial among aquarists. Some have called blood parrot cichlids "the Frankenstein monster of the fish world." Another notable hybrid, the flowerhorn cichlid, was very popular in some parts of Asia from 2001 until late 2003, and is believed to bring good luck to its owner. The popularity of the flowerhorn cichlid declined in 2004. Owners released many specimens into the rivers and canals of Malaysia and Singapore where they threaten endemic communities.

Numerous cichlid species have been selectively bred to develop ornamental aquarium strains. The most intensive programs have involved angelfish and discus, and many mutations that affect both colouration and finnage are known. Other cichlids have been bred for albino, leucistic, and xanthistic pigment mutations, including oscars, convicts and *Pelvicachromis pulcher*. Both dominant and recessive pigment mutations have been observed. In convict cichlids, for example, a leucistic colouration is recessively inherited, while in *Oreochromis niloticus niloticus* red colouration is caused by a dominant inherited mutation.

This selective breeding may have unintended consequences. For example, hybrid strains of *Mikrogeophagus ramirezi* have health and fertility problems. Similarly, intentional inbreeding can cause physical abnormalities, such as the notched phenotype in angelfish.

5

Fresh Water Ornamental Fish Farming

Fishing Industry in the United States

As with other countries, the 200 nautical miles (370 km) exclusive economic zone (EEZ) off the coast of the United States gives its fishing industry special fishing rights. It covers 11.4 million square kilometres (4.38 million sq mi). This is the largest zone in the world, exceeding the land area of the United States.

According to the FAO, in 2005 the United States harvested 4,888,621 tonnes of fish from wild fisheries and another 471,958 tonnes from aquaculture. This made the United States the fifth leading producer of fish after China, Peru, India and Indonesia, with 3.8 percent of the world total.

Management

Historically, fisheries developed in the U.S. as each area was settled, whether by the original aboriginal peoples or by post Columbian arrivals. Concern for the sustainability of fishery resources was evident as early as 1871, when Congress wrote that "... the most valuable food fishes of the coast and the lakes of the U.S. are rapidly diminishing in number, to the public injury, and so as materially to affect the interests of trade and commerce...." However, it was not until 1976, with the Magnuson-Stevens Fishery Conservation and Management Act, that the federal government began actively managing fisheries.

Today, inland fisheries and nearshore marine fisheries are managed by state (or regional or county) fisheries commissions. State jurisdictions usually extend 3 nautical miles (6 km) out to sea. Coastal fisheries in the EEZ beyond state jurisdictions are the responsibility of the federal system. The primary institutions of the federal system are eight regional fishery management councils and the National Marine Fisheries Service (NMFS), also known as NOAA Fisheries. NMFS works in partnership with industry, universities, and state, local, and tribal agencies to collect data about commercial species.

Fisheries observers on fishing vessels, transmit real time data electronically to NMFS.

NMFS works in partnership with the regional fishery management councils to prevent overfishing and restore overfished stocks. Objectives are to reduce fishing intensity, monitor the fisheries, and implement measures to reduce by catch and protect essential fish habitat. NMFS is establishing marine protected areas and individual fishing quotas, and implementing ecosystem based fishery management.

Wild Fisheries

EEZ

The U.S. EEZ is the largest in the world, 1.7 times the land area of the US, and includes eight large marine ecosystems:

- Northeast U.S. Continental Shelf
- Southeast U.S. Continental Shelf
- Caribbean Sea
- Gulf of Mexico
- California Current
- Insular Pacific-Hawaiian
- Gulf of Alaska.

Nearshore Fisheries

Nearshore fishing areas consist of estuaries and coastal areas that lie within 3 nautical miles (6 km) of the shoreline. These are under the control of the relevant coastal state rather than under federal or NMFS control. They vary widely in species diversity and abundance. Many species are highly prized game fish, while others are small forage fishused for bait, animal food, and industrial products. Those of greatest interest include invertebrate species like crabs, shrimps, abalones, clams, oysters and scallops. Many species are of unknown status because it is difficult to assess their condition. There are no firm estimates exist for long term potential yield.

Management is spread out among the coastal states and other local authorities, and a comprehensive treatment of the fisheries has not been attempted. Traditional techniques are usually employed, including size limits, catch limits, method restrictions, and area and time closures.

Coastal Fisheries

The coastal fisheries are the fisheries that lie within 200 nautical miles (370 km) of the coast that defines the exclusive economic zone, but outside the 3 nautical miles (6 km) distance that defines the nearshore fisheries. Coastal fisheries are under federal control in the form of NMFS, as well as under the control of one of the eight regional fishery management councils

Northeast

Traditionally the most valued fishery for the northeast region has been groundfish. However, the groundfish, especially haddock, yellow tail flounder and cod have been overfished. Record low spawning biomass levels occurred in 1993–94, though these are now recovering. Dogfish and skate rebounded in the 1970s while groundfish and flounder declined. These fish are an important part of the Georges Bank.

The next most important fishery by value is American lobster and Atlantic sea scallop. The striped bass was driven to low levels early in the 1980s. Catch restrictions were applied in the mid 1980s, and by 1995 this an adromous fish had recovered. The region has valuable mollusk fisheries. Offshore are sea scallops, surfclams, American lobsters and ocean quahog. Inshore are oysters, blue mussels, blue crabs, and clam fisheries. These fisheries are pretty much fully exploited.

The fisheries in the Northeast Region are governed mostly by Fishery Management Plans (FMPs). An example of successful regulation is the 1994 Amendment 4 to the Sea Scallop FMP. This set out to control scallop fishing effort by not accepting more entrants, by restricting the time vessels could fish each day, and by requiring bigger mesh sizes on dredges. Additional protection for scallops was given closing parts of Georges Bank to groundfish fishing. The scallops recovered.

Southeast

The Southeast region spans the Gulf of Mexico, the Caribbean Sea and the US Southeast Atlantic. Important species are menhaden, drum, croaker, invertebrates, highly migratory species, reef fish and other nearshore species.

Overfishing of king and Spanish mackerel occurred in the 1980s. Regulations were introduced to restrict the size, amount of catch, fishing locations and bag limits for recreational fishers as well as commercial fishers. Gillnets were banned in waters off Florida. By 2001, the mackerel stocks had bounced back.

Menhaden and stone crabs and the three major species of shrimp (pink, brown and white) are fully exploited. Spiny lobster are overexploited. Better data is needed if stocks are to be assessed accurately.

Alaska

The salmon species in Alaska (chinook, coho, pink, sockeye, and chum) generally produce good harvest, though some stocks are declining. Up to 1977, foreign fishing controlled the groundfish fisheries in Alaska (apart from Pacific halibut).

The Aleutian Islands, a series of over 300 rocky islands, stretch over 1,000 miles (1,600 km) from southwest Alaska to Russia. They are home to the largest fishing port in the U.S., Dutch Harbour. The primary target species is pollock, but crabs, salmon and groundfish are also important. Between 1997 and 2001, groundfish were caught bybottom trawling, a fishing method which destroys the habitat of the groundfish, the ocean floor coral and sponge communities. In 2005, about 80,000 km^2 (30,000 sq mi) of seafloor around the Aleutian Islands were permanently closed to destructive fishing gears. In 2008, NOAA fisheries created the Aleutian Islands Habitat Conservation Area (AIHCA), most of which is closed to bottom trawling. At 725,000 km^2 (280,000 sq mi), the AIHCA is the largest marine protected area in the USA.

Pacific Coast

On the Pacific Coast important species are Pacific halibut, Pacific salmon, groundfish, pelagic fishes and nearshore species. The stocks are mostly overfished or fully exploited.

- Salmon: Salmon production has decreased since the late 1970s, partly due to habitat degradation.
- Groundfish: this harvest is dominated by Pacific whiting. Some stocks need rebuilding. Rockfish can live as long as 100 years, and grow and reproduce very slowly. This makes stock recovery very slow. In 2004, plans for rebuilding several overfished groundfish species were approved .
- Shellfish: Shellfish, such as crabs, clams, shrimp and abalone, sell for high prices, so their fisheries can be small by volume but high by value. These fisheries are mostly fully exploited recreational fishers catch more of some species than the commercial fishers.
- Recovering stocks: After decades of decline, Pacific sardine populations are recovering. In 2004, the NMFS declared that, due to reduced harvest and bycatch, Pacific whiting were recovering.

Western Pacific

The Western Pacific region includes the western and central Pacific, the Hawaiian Islands, and the islands of the Northern Marianas, Guam and American Samoa. These are tropical or subtropical waters. They have large species diversity but, because ocean nutrients are not rich, the sustainable yields is low. Pelagic armourhead is the only overfished stock.

- Groundfish fisheries: Groundfish, such as snapper jacks, emperors and grouper are harvested from coral and rock habitats, mostly around the Hawaiian Islands. While some stocks are underutilised, other important species have declined to 30 percent of their earlier stock levels.
- Invertebrate fisheries: The main invertebrate fisheries are in the north west Hawaiian Islands for slipper and spiny lobster. This fishery started in 1977, peaked in the mid 1980s, and then declined. The decline is thought to stem from oceanographic changes. Some recovery has occurred since 1991, due to entry and harvest restrictions.
- Highly migratory species account for 99 percent catch in this region and are discussed below.

High Sea Fisheries

The high seas, or international waters, are the waters outside the jurisdiction of the EEZ of any country. U. S. distant fisheries operate in some of these waters, alongside other nations, beyond the jurisdiction of any nation's coastal fisheries.

Highly Migratory Species

The main commercial species in the high seas are the highly migratory species. These fish make long migrations across the high seas and are fished by many nations. Highly migratory fish also cross boundaries without regard for international laws. In particular, they enter the EEZ zones of the U.S., which means they become important species also for U.S. coastal fisheries.

In the Atlantic, overfished species include:

- Tuna: bigeye in the Atlantic, albacore in the North Atlantic, and bluefin in the West Atlantic, with yellowfin tuna close to being overfished.
- Blue marlin, white marlin and sailfish.
- Sharks: bull, dusky, bignose, night, Caribbean reef, spinner, silky, tiger, lemon, sand, nurse, scalloped, smooth smooth hammerhead and white shark.

The biomass of swordfish in the North Atlantic has increased, probably due to catch reduction and strong recruitment.

In the Pacific, these transboundary fisheries are also important to other Pacific Rim nations and to U.S. fleets fishing within and beyond the EEZ. The major U.S. catch stock is tuna, though billfish, swordfish and shark are also caught. These stocks account for 99 percent of the Western Pacific region's catch.

Seamounts

A seamount is a mountain rising from the ocean seafloor that does not reach the ocean surface. They are defined by oceanographers as independent features that rise at least 1,000 metres above the seafloor. Seamounts became interesting during the 1960s when it was discovered that they can maintain large stocks of commercially important fishes and invertebrates.

From 1968 to about 1990, foreign fleets harvested pelagicarmourhead across seamounts in the Hawaiian Ridge. Since 1984, fishing been prohibited there so the stock can recover.

Generally, seamounts are found in the high seas, outside the continental shelves and outside the jurisdiction of any nation's EEZ. However, the U.S. Pacific Coast region has an unusually small continental shelf, while the Western Coast region contains islands with no continental shelf. As a result, there are some seamounts within U.S. jurisdiction.

Inland Fisheries

Fisheries in inland waters of the United States are small compared to marine fisheries. The largest fisheries are the landings from the Great Lakes, worth about $13 million in 2003, with a similar amount from the Mississippi River basin. This is less than one percent of the dollar value of the marine fisheries.

Fishing Fleet

Until the late 19th century, the U.S. fishing fleet used sailing vessels. By the early 20th century, fishing vessels were built as steam boats with steam engines, or as schooners with auxiliary gasoline engines. By the 1930s the fleet was almost completely converted to diesel vessels.

Fishing gear became more technical: Alaska purse seiners were in use by 1870, longliners were introduced in 1885; otter trawls were operating in the groundfish and shrimp fisheries by the early 20th century. In the late 1960s, factory ships from other countries started fishing haddock, herring, salmon and halibut on traditional U.S. fishing grounds.

Technological advances have played an important role in the development of U.S. fisheries. Increases in size and speed allowed vessels to fish in more distant waters.

Advances include double trawls, the Puretic power blocks for retrieving seine nets, refrigerated holds, durable synthetic fibres for lines and nets, GPS to navigate and locate fishing grounds, fishfinders for the location of fish, and spotter planes to locate fish schools.

The vessel size and gear types of the U.S. fishing fleet varies geographically and between fisheries. Because of this, management of U.S. fisheries with a single policy is not feasible. Individual fisheries have their own biological, economic, and sociological characteristics which make broad policies impractical. On the other hand, ad hoc regulations for individual fisheries is not practical either.

U.S. fisheries use most fishing gear types. Vessels are often configured so they can change rapidly between two or more gear types, such as lobster pots to bottom trawls to scallop dredges. The main techniques are purse seining and trawling. Some vessels freeze their catch at sea, such as factory trawlers, tuna boats, Alaskan crab pot vessels, and some southeast shrimp trawlers. Vessels usually land their catches near their homeports.

Based in Hawaii is a longline fishery for tuna and billfish. New management approaches, including placing fisheries observers on vessels to observe what is happening with protected species such as albatross and leatherback, loggerhead and green sea turtles, is expected to reduce incidental catch.

Overfished Stocks

By 2003, overfishing had occurred on 60 stocks. Another 232 stocks were not overfished. Overfishing had been stopped on 31 stocks, and a gain was made of 13 stocks that had been fully rebuilt.

There were 617 other stocks which have limited data or for which criteria for overfishing had not been defined. These stocks mostly have harvests which are not significant, so they are not allocated research funding. Their assessment can be expensive and there is no evidence of overfishing. Rebuilding strategies are operating or are being developed for most stocks which are overfished.

There were 267 major stocks, that is, stocks with at least 200,000 pounds (91,000 kg) in annual landings. Of these, 40 were overfished, 147 were not, and it was not known whether the remaining 80 stocks were overfished. There were 642 minor stocks, of which 20 were overfished, 85 were not, and the remainder had an overfishing status which is unknown or is undefined.

Marine Protected Areas

A marine protected area (MPA) is federally defined as: "any area of the marine environment that has been reserved by federal, state, tribal, territorial, or local laws or regulations to provide lasting protection for part or all of the natural and cultural resources therein."

Fourteen MPAs covering 150,000 square miles (390,000 km^2) have been federally-designated as National Marine Sanctuaries, and are administered by a division of the National Oceanic and Atmospheric Administration (NOAA).

In practice, these MPAs are defined areas where natural and/or cultural resources are given more protection than occurs in the surrounding waters. United States MPAs cover many habitats including the open ocean, coastal areas, intertidal zones, estuaries, and the Great Lakes. They vary in purpose, legal authorities, agencies, management approaches, level of protection, and restrictions on human uses.

Aquaculture

The value of aquaculture products grew from $45 million in 1974 to about $866 million (393,400 tonnes) in 2003. Aquaculture, in the United States, includes the farming of hatchery fish and shellfish which are grown to market size in ponds, tanks, cages, or raceways and released into the wild. Aquaculture is also used to support commercial and recreational marine fisheries by enhancing or rebuilding wild stock populations. It also includes the cultivation of ornamental fish for the aquarium trade, as well as plant species used in various pharmaceutical, nutritional, and biotechnology products.

According to the FAO, in 2004 the United States ranked 10th in total aquaculture production, behind China, India, Vietnam, Thailand, Indonesia, Bangladesh, Japan, Chile, and Norway. The United States imports aquaculture products from these and other countries, and operates an annual seafood trade deficit of over $9 billion.

Shellfish (oysters, clams, mussels), account for two-thirds of marine aquaculture production, followed by salmon (25 percent) and shrimp (10 percent). Production occurs mainly on land, in ponds, and in coastal waters under state jurisdiction. As a result of recent advances in offshore aquaculture technology, commercial finfish and shellfish operations have been established in more exposed, open ocean locations in state waters in New Hampshire, Hawaii, and Puerto Rico.

Total U.S. aquaculture production, including aquatic plants, is about $1 billion annually, compared to the total world production of about $70 billion. Only about 20 percent of U.S. aquaculture production is from marine species. NOAA estimates that the annual U.S. domestic aquaculture production of all species could increase from about 0.5 million tons to 1.5 million tons by 2025.

Subsidies

In a study of federal investment in the fishery sector, most aspects of United States tax, fisheries, and societal policies were examined to see whether they created subsidies for the fishing industry and whether these subsidies had impacts. The gross value of direct U.S. subsidies was about 0.5% of the gross ex-vessel value of commercial landings. There are no major ship construction subsidies, market development or other forms of assistance that often exist in developed and developing fishing industries around the world.

Recreational

In 2006, the U.S. Fish and Wildlife Service estimated that 30.0 million U.S. anglers, 16 years old and older took 403 million fishing trips, spending $42.0 billion in fishing related expenses. Of these, 25.4 million were freshwater anglers who took 337 million trips and spent $26.3 billion. Saltwater fishing attracted 7.7 million anglers who took 67 million trips and spent $8.9 billion.

Recreational fishing expenditures in $US billion			
	subtotal	***total***	***percent***
Food and lodging	6.3		
Transportation	5.0		
Other trip costs: land use fees, guide fees, equipment rental, boating expenses, and bait...	6.6		
Trip-related, total		17.9	
Fishing equipment: rods, reels, tackle boxes, depth finders, and artificial lures and flies...	5.3		
Auxiliary equipment: camping equipment, binoculars, and special fishing clothing...	0.8		
Special equipment: boats, vans, and cabins...	13.2		
Equipment, total		18.8	
Magazines, books	0.1		
Membership dues and contributions	0.2		
Land leasing and ownership	4.6		
Licenses, stamps, tags, and permits	0.5		
Other, total		5.4	
Overall total		42.0	100

Magnuson–Stevens Fishery Conservation and Management Act

The Magnuson–Stevens Fishery Conservation and Management Act (MFCMA), commonly referred to as the Magnuson–Stevens Act, is the primary law governing marine fisheries management in the United States. The law is named after Warren G. Magnuson, former U.S. senator from Washington state, and Ted Stevens, the former senator from Alaska.

The Magnuson–Stevens Act was originally enacted as the *Fishery Conservation and Management Act of 1976*. The act has been amended many times over the years. Two major recent sets of amendments to the law were the Sustainable Fisheries Act of 1996 and then 10 years later the *Magnuson–Stevens Fishery Conservation and Management Reauthorisation Act of 2006*.

Purpose

The MFCMA was enacted to promote the U.S. fishing industry's optimal exploitation of coastal fisheries by "consolidating control over territorial waters" and establishing eight regional councils to manage fish stocks. The act has been amended several times in response to continued overfishing of major stocks. The most recent version, authorised in 2007, includes seven purposes:

1. Acting to conserve fishery resources
2. Supporting enforcement of international fishing agreements
3. Promoting fishing in line with conservation principles
4. Providing for the implementation of fishery management plans (FMPs) which achieve optimal yield

5. Establishing Regional Fishery Management Councils to steward fishery resources through the preparation, monitoring, and revising of plans which (A) enable stake holders to participate in the administration of fisheries and (B) consider social and economic needs of states.
6. Developing underutilised fisheries
7. Protecting essential fish habitats

Additionally, the law calls for reducing bycatch and establishing fishery information monitoring systems.

Regulatory Mechanisms

Regional Fishery Management Councils are charged with developing and implementing fishery management plans, both to restore depleted stocks and manage healthy stocks. The National Marine Fisheries Service (NMFS) aides the Secretary of Commerce, who evaluates and approves the council's FMPs. Regional Fishery Management Council members are nominated by the governors of their respective states, and approved by the Secretary of Commerce.

A FMP must specify the criteria which determine when a stock is overfished and the measures needed to rebuild it. Regional councils regulate fishers with mechanisms, including annual catch limits, individual catch limits, community development quotas, and others. The Marine Fish Conservation Network highlighted the most significant changes in the mechanisms utilised in a 2010 report:

> *To achieve the goal of ending overfishing ... Congress strengthened the role of science in the fishery management process and required fishery managers to establish science based annual catch limits (ACLs) and accountability measures (AMs) for all US fisheries with a deadline of 2010 for all stocks subject to overfishing... The new fisheries law requires the councils' science advisors, the scientific and statistical committees to make recommendations for 'acceptable biological catch' (ABC) which managers may not exceed...*

The ACL is the centrepiece of the report which is supplemented by other mechanisms regulating the types of gear used, licensing vessels, and using of observers on fishing boats. In section 303 b, the Act enumerates the types of actions authorised for use by councils to achieve optimal catch goals. Including

1. Permitting vessels or operators
2. Designating Zones and periods where fishing is limited
3. Limiting sale, catch or transport of certain fish.
4. Regulating types of fishing equipment
5. Requiring observers onboard vessels.

Regional management councils have taken the controversial step of buying fishing vessels to remove them from the water to reduce the size of the fishing fleet.

Regulatory Effectiveness

The act's results vary for different regions and different fish stocks. It did not prevent the overfishing of many species throughout its first 20 years of existence. This prompted major amendments in 1996 and 2006. The National Marine Fisheries Service issued a report to Congress in 2010 on the status of U.S. fisheries. It reported that of the 192 stocks monitored for overfishing 38 stocks (20%) still have fish "mortality rates that exceed the overfishing threshold ... and 42 stocks (22%) are overfished". This is down from 38% and 48% respectively in 2000. A 2003 NMFS report reviewed achievements of the act since 1996. Highlighting the inconsistent effects of the legislation, it revealed that overfishing was eliminated in 15 major fish stocks while overfishing was initiated in 12 major fish stocks. To improve their overfishing prevention programs, the NMFS has implemented the Fish Stock Sustainability Index (FSSI), which measures key stocks according to their overfishing status and biomass levels. Since the FSSI began the index has increased every year.

The act also has impacts on financial matters. While taxpayers have paid over US$3 billion on NFMS programs since the bills inception, ultimately the fishing industry is most affected by the act's design flaws and incomplete implementation. According to Zeke Grader, Jr., the executive director of the Pacific Coast Federation of Fishermen's Associations, the largest active trade association of commercial fishermen on the west coast, "Most of the U.S. fisheries stocks are facing a disaster due to over capitalisation of the fishing industry and the mismanagement practices of U.S. Department of Commerce's National Marine Fisheries Service (NMFS) and their appointed regional fishery management councils". Chris Kellogg, chief technical officer for the New England Fishery Management Council emphasizes the tenuous position of many fishers, "On average, my guess is the fishing harvesting industry here in New England is basically covering costs and just on the border of solvency and insolvency".

Stakeholders

Many interest groups are concerned with the forming of fisheries conservation legislation. Fishers, corporations, activist groups and the public all share interest in protecting fishing eco-systems and economies via the MSFCMA.

Fishers advocate measures that encourage regulatory processes to be scaled to the local level and which ensure fishing privileges aren't concentrated into small groups. They are also aware that if too much competition for finite resources prevails, their livelihood will suffer. Despite this, some fishermen prefer minimal government intervention in their market, defiantly demanding "the right to go broke".

The fishing industry is a $4billion a year industry. Fish harvesting and processing corporations are invested in the political process to maximise their profits, to protect against foreign competition and to prevent regulations from making their proprietary information available to the public. University, argues that federal legislators cannot forge these relationships largely because they haven't properly identified what a "fishing community" is. She highlights the successes of the Community Development Quota system employed in some Alaska fisheries.

Non-Governmental Organisations (NGOs) focusing on fish conservation and environmentalism were key forces behind the incremental improvements in the law's regulatory mechanisms. The Pew Oceans Commission and the US commission on Oceans Policy prompted many of the amendments found within the 2006 reauthorisation. Many advocacy groups speak through coalitions. The Marine Fish Conservation Network, for example, represents over 90 member organisations from across the United States.

The public is represented as a stake holder by elected representatives, who ostensibly take them into consideration when drafting ways to protect public resources such as fish stocks.

Critiques

The major criticisms of the act have been its failure to stem overfishing, minimise bycatch, and to hold accountable regional councils that don't enforce or implement fisheries management plans. Additionally, the act is accused of coddling fishers who make poor business decisions, inadequately protecting ecosystems, and lacking transparency requirements for fishery related data.

Poor enforcement of previous versions of the act is often blamed for depleting stocks. Fishermen have accused each other of cheating on landings and chastised regulators for concentrating the quota allocations into too few hands. Other critics claim the regulatory framework is too "top-down" and alienates local fishers, thereby reducing the likelihood to achieve the cooperation needed to enforce many provisions. Candice May, of Colorado State University, argues that federal legislators can not forge these relationships largely because they haven't properly identified what a "fishing community" is. She highlights the successes of the Community Development Quota system employed in some Alaska fisheries.

In response to these types of criticisms, "On May 1, 2010, the National Marine Fisheries Service (NMFS) implemented a new management system for ground fish in New England. It established 17 fishermen-run collectives, called sectors. Sectors were pioneered by fishermen as voluntary, cooperative and community-based, and were designed to protect fleet diversity and coastal communities. The new management system operates on three simple premises:

- It implements science-based catch limits to rebuild fish populations and prevent overfishing.
- It incorporates monitoring so fishermen and regulators know exactly how much fish is being caught, and as a result, fishing stops once catch limits have been reached.
- Each sector receives its own share of the annual catch. While respecting catch limits, the co-ops provide fishermen with the flexibility to set their own fishing guidelines so they can run their businesses more efficiently and profitably. Those who develop more innovative fishing gear can target more of the healthy fish populations and avoid those populations that are struggling."

As reported at the Council meeting, the first three months of sector operations resulted in (May 1 – August 15):

- Fishermen earning more money for less fishing under the new system. In 2010, landings are down compared to 2009. Only 85.8 percent of total landings last year were landed this year (for the same period of time). Meanwhile, revenues are up 112.4 percent.
- Sector fishermen are avoiding weak stocks and targeting robust stocks. The ratio of Georges Bank cod to Georges Bank haddock (in metric tons) in 2009 was 1121:1532. In 2010, it was 743:2768.
- Landings of Gulf of Maine winter flounder, a stock at very low abundance, are being effectively avoided under sectors. In 2009, 66 metric tons were landed. In 2010, 32 metric tons were landed."

Fisheries Management

Fisheries management draws on fisheries science in order to find ways to protect fishery resources so sustainable exploitation is possible. Modern fisheries management is often referred to as a governmental system of appropriate management rules based on defined objectives and a mix of management means to implement the rules, which are put in place by a system of monitoring control and surveillance. According to the FAO, there are "no clear and generally accepted definitions of fisheries management". However, the working definition used by the FAO and much quoted elsewhere is:

> *The integrated process of information gathering, analysis, planning, consultation, decision-making, allocation of resources and formulation and implementation, with enforcement as necessary, of regulations or rules which govern fisheries activities in order to ensure the*

continued productivity of the resources and the accomplishment of other fisheries objectives.

History

Fisheries have been explicitly managed in some places for hundreds of years. For example, the Mâori people, New Zealand residents for about 700 years, had prohibitions against taking more than could be eaten and about giving back the first fish caught as an offering to sea god Tangaroa. Starting in the 18th century attempts were made to regulate fishing in the North Norwegian fishery. This resulted in the enactment of a law in 1816 on the Lofoten fishery, which established in some measure what has come to be known as territorial use rights.

> *"The fishing banks were divided into areas belonging to the nearest fishing base on land and further subdivided into fields where the boats were allowed to fish. The allocation of the fishing fields was in the hands of local governing committees, usually headed by the owner of the onshore facilities which the fishermen had to rent for accommodation and for drying the fish."*

Governmental resource protection-based fisheries management is a relatively new idea, first developed for North European fisheries after the first Overfishing Conference held in London in 1936. In 1957 British fisheries researchers Ray Beverton and Sidney Holt published a seminal work on North Sea commercial fisheries dynamics. In the 1960s the work became the theoretical platform for North European management schemes.

After some years away from the field of fisheries management, Beverton criticized his earlier work in a paper given at the first World Fisheries Congress in Athens in 1992. "The Dynamics of Exploited Fish Populations" expressed his concerns, including the way his and Sydney Holt's work had been misinterpreted and misused by fishery biologists and managers during the previous 30 years. Nevertheless, the institutional foundation for modern fishery management had been laid.

Political Objectives

According to the FAO, fisheries management should be based explicitly on political objectives, ideally with transparent priorities. Typical political objectives when exploiting a fish resource are to:

- maximise sustainable biomass yield
- maximise sustainable economic yield
- secure and increase employment
- secure protein production and food supplies
- increase export income

Such political goals can also be a weak part of fisheries management, since the objectives can conflict with each other.

International Objectives

Fisheries objectives need to be expressed in concrete management rules. In most countries fisheries management rules should be based on the internationally agreed, though non-binding, Code of Conduct for Responsible Fisheries, agreed at a meeting of the U.N.'s Food and Agriculture Organisation FAO session in 1995. The precautionary approach it prescribes is typically implemented in concrete management rules as minimum spawning biomass, maximum fishing mortality rates, etc. In 2005 the Fisheries Centre at the University of British Columbia comprehensively reviewed the performance of the world's major fishing nations against the Code.

International agreements are required in order to regulate fisheries in international waters. The desire for agreement on this and other maritime issues led to three conferences on the Law of the Sea, and ultimately to the treaty known as the United Nations Convention on the Law of the Sea (UNCLOS). Concepts such as exclusive economic zones (EEZ, extending 200 nautical miles (370 km) from a nation's coasts) allocate certain sovereign rights and responsibilities for resource management to individual countries.

Other situations need additional intergovernmental coordination. For example, in the Merranean Sea and other relatively narrow bodies of water, EEZ of 200 nautical miles (370 km) are irrelevant. International waters beyond 12-nautical-mile (22 km) from shore require explicit agreements.

Straddling fish stocks, which migrate through more than one EEZ also present challenges. Here sovereign responsibility must be agreed with neighbouring coastal states and fishing entities. Usually this is done through the medium of a regional organisation set up for the purpose of coordinating the management of that stock.

UNCLOS does not prescribe precisely how fisheries confined only to international waters should be managed. Several new fisheries (such as high seas bottom trawling fisheries) are not (yet) subject to international agreement across their entire range. In November 2004 the UN General Assembly issued a resolution on Fisheries that prepared for further development of international fisheries management law.

Management Mechanisms

Many countries have set up Ministries/Government Departments, named "Ministry of Fisheries" or similar, controlling aspects of fisheries within their exclusive economic zones. Four categories of management means have been devised, regulating either input/investment, or output, and operating either directly or indirectly:

	Inputs	***Outputs***
Indirect	Vessel licensing	Catching techniques
Direct	Limited entry	Catch quota and technical regulation

Technical means may include:

- prohibiting devices such as bows and arrows, and spears, or firearms
- prohibiting nets
- setting minimum mesh sizes
- limiting the average potential catch of a vessel in the fleet (vessel and crew size, gear, electronic gear and other physical "inputs".
- prohibiting bait
- snagging
- limits on fish traps
- limiting the number of poles or lines per fisherman
- restricting the number of simultaneous fishing vessels
- limiting a vessel's average operational intensity per unit time at sea
- limiting average time at sea.

Catch Quotas

Individual Transferable Quota: Systems that use *individual transferable quotas* (ITQ), also called individual fishing quota limit the total catch and allocate shares of that quota among the fishers who work that fishery. Fishers can buy/sell/trade shares as they choose.

A large scale study in 2008 provided strong evidence that ITQ's can help to prevent fishery collapse and even restore fisheries that appear to be in decline.

Precautionary Principle

The *Fishery Manager's Guidebook* issued in 2009 by the FAO of the United Nations, advises that the precautionary approach or principle should be applied when "ecosystem resilience and human impact (including reversibility) are difficult to forecast and hard to distinguish from natural changes." The precautionary principle suggests that when an action risks harm, it should not be proceeded with until it can be scientifically proven to be safe. Historically fishery managers have applied this principle the other way round; fishing activities have not been curtailed until it has been proven that they have already damaged existing ecosystems. In a paper published in 2007, Shertzer and Prager suggested that there can be significant benefits to stock biomass and fishery yield if management is stricter and more prompt.

Climate Change

In the past, changing climate has affected inland and offshore fisheries and such changes are likely to continue. From a fisheries perspective, the specific driving factors of climate change include rising water temperature, alterations in the hydrologic cycle, changes in nutrient fluxes, and relocation of spawning and nursery habitat. Further, changes in such factors would affect resources at all levels of biological organisation, including the genetic, organism, population, and ecosystem levels.

Population Dynamics

Population dynamics describes the growth and decline of a given fishery stock over time, as controlled by birth, death and migration. It is the basis for understanding changing fishery patterns and issues such as habitat destruction, predation and optimal harvesting rates. The population dynamics of fisheries has been traditionally used by fisheries scientists to determine sustainable yields.

The basic accounting relation for population dynamics is the BIDE model:

$$N_1 = N_0 + B - D + I - E$$

where N_1 is the number of individuals at time 1, N_0 is the number of individuals at time 0, B is the number of individuals born, D the number that died, I the number that immigrated, and E the number that emigrated between time 0 and time 1. While immigration and emigration can be present in wild fisheries, they are usually not measured.

Care is needed when applying population dynamics to real world fisheries. In the past, over-simplistic modelling, such as ignoring the size, age and reproductive status of the fish, focusing solely on a single species, ignoring bycatch and physical damage to the ecosystem, has accelerated the collapse of key stocks.

Ecosystem-Based Fisheries

We propose that rebuilding ecosystems, and not sustainability per se, should be the goal of fishery management. Sustainability is a deceptive goal because human harvesting of fish leads to a progressive simplification of ecosystems in favour of smaller, high turnover, lower trophic level fish species that are adapted to withstand disturbance and habitat degradation. —Tony Pitcher and Daniel Pauly,

Ecosystem-based management: According to marine ecologist Chris Frid, the fishing industry points to pollution and global warming as the causes of unprecedentedly low fish stocks in recent years, writing, "Everybody would like to see the rebuilding of fish stocks and this can only be achieved if we understand all of the influences, human and natural, on fish dynamics." Overfishing has also had an effect. Frid adds, "Fish communities can be altered in a number of ways, for example they can decrease if particular sized individuals of a species are targeted, as this affects predator and prey dynamics. Fishing, however, is not the sole perpetrator of changes to marine life - pollution is another example ... No one factor operates in isolation and components of the ecosystem respond differently to each individual factor."

In contrast to the traditional approach of focusing on a single species, the ecosystem-based approach is organised in terms of ecosystem services. Ecosystem-based fishery concepts have been implemented in some regions. In 2007 a group of scientists offered the following *ten commandments*

- Keep a perspective that is holistic, risk-adverse and adaptive.
- Maintain an "old growth" structure in fish populations, since big, old and fat female fish have been shown to be the best spawners, but are also susceptible to overfishing.
- Characterise and maintain the natural spatial structure of fish stocks, so that management boundaries match natural boundaries in the sea.
- Monitor and maintain seafloor habitats to make sure fish have food and shelter.
- Maintain resilient ecosystems that are able to withstand occasional shocks.
- Identify and maintain critical food-web connections, including predators and forage species.
- Adapt to ecosystem changes through time, both short-term and on longer cycles of decades or centuries, including global climate change.
- Account for evolutionary changes caused by fishing, which tends to remove large, older fish.
- Include the actions of humans and their social and economic systems in all ecological equations.
- Report to Congress (2009): The State of Science to Support an Ecosystem Approach to Regional Fishery Management National Marine Fisheries Service, NOAA Technical Memorandum NMFS-F/SPO-96.

Elderly Maternal Fish

Traditional management practices aim to reduce the number of old, slow-growing fish, leaving more room and resources for younger, faster-growing fish. Most marine fish produce huge numbers of eggs. The assumption was that younger spawners would produce plenty of viable larvae.

However, 2005 research on rockfish shows that large, elderly females are far more important than younger fish in maintaining productive fisheries. The larvae produced by these older maternal fish grow faster, survive starvation better, and are much more likely to survive than the offspring of younger fish. Failure to account for the role of older fish may help explain recent collapses of some major US West Coast fisheries. Recovery of some stocks is expected to take decades. One way to prevent such collapses is to establish marine reserves, where fishing is not allowed and fish populations age naturally.

Data Quality

According to fisheries scientist Milo Adkison, the primary limitation in fisheries management decisions is the absence of quality data. Fisheries management decisions are often based on population models, but the models need quality data to be effective. He asserts that scientists and fishery managers would be better served with simpler models and improved data.

The most reliable source for summary statistics is the FAO Fisheries Department.

Ecopath

Ecopath, with Ecosim (EwE), is an ecosystem modelling software suite. It was initially a NOAA initiative led by Jeffrey Polovina, later primarily developed at the Fisheries Centre of the University of British Columbia. In 2007, it was named as one of the ten biggest scientific breakthroughs in NOAA's 200-year history. The citation states that Ecopath "revolutionised scientists' ability worldwide to understand complex marine ecosystems". Behind this lies two decades of development work by Villy Christensen, Carl Walters, Daniel Pauly, and other fisheries scientists. As of 2010 there are 6000 registered users in 155 countries. Ecopath is widely used in fisheries management as a tool for modelling and visualising the complex relationships that exist in real world marine ecosystems.

Human Factors

Managing fisheries is about managing people and businesses, and not about managing fish. Fish populations are managed by regulating the actions of people. If fisheries management is to be successful, then associated human factors, such as the reactions of fishermen, are of key importance, and need to be understood.

Management regulations must also consider the implications for stakeholders. Commercial fishermen rely on catches to provide for their families just as farmers rely on crops. Commercial fishing can be a traditional trade passed down from generation to generation. Most commercial fishing is based in towns built around the fishing industry; regulation changes can impact an entire town's economy. Cuts in harvest quotas can have adverse effects on the ability of fishermen to compete with the tourism industry.

Performance

The biomass of global fish stocks has been allowed to run down. This biomass is now diminished to the point where it is no longer possible to sustainably catch the amount of fish that could be caught. According to a 2008 UN report, titled *The Sunken Billions: The Economic Justification for Fisheries Reform*, the world's fishing fleets incur a "$US 50 billion annual economic loss" through depleted stocks and poor fisheries management.

The report, produced jointly by the World Bank and the UN Food and Agriculture Organisation (FAO), asserts that half the world's fishing fleet could be scrapped with no change in catch.

> *By improving governance of marine fisheries, society could capture a substantial part of this $50 billion annual economic loss. Through comprehensive reform, the fisheries sector could become a basis for economic growth and the creation of alternative livelihoods in many countries. At the same time, a nation's natural capital in the form of fish stocks could be greatly increased and the negative impacts of the fisheries on the marine environment reduced.*

The most prominent failure of fisheries management in recent times has perhaps been the events that lead to the collapse of the northern cod fisheries. More recently, the International Consortium of Investigative Journalists produced a series of journalistic investigations called *Looting the seas*. These detail investigations into the black market for bluefin tuna, the subsidies propping up the Spanish fishing industry, and the overfishing of the Chilean jack mackerel.

Sustainable Fishery

A conventional idea of a sustainable fishery is that it is one that is harvested at a sustainable rate, where the fish population does not decline over time because of fishing practices.

Sustainability in fisheries combines theoretical disciplines, such as the population dynamics of fisheries, with practical strategies, such as avoiding overfishing through techniques such as individual fishing quotas, curtailing destructive and illegal fishing practices by lobbying for appropriate law and policy, setting up protected areas, restoring collapsed fisheries, incorporating all externalities involved in harvesting marine ecosystems into fishery economics, educating stakeholders and the wider public, and developing independent certification programs.

Some primary concerns around sustainability are that heavy fishing pressures, such as overexploitation and growth or recruitment overfishing, will result in the loss of significant potential yield; that stock structure will erode to the point where it loses diversity and resilience to environmental fluctuations; that ecosystems and their economic infrastructures will cycle between collapse and recovery; with each cycle less productive than its predecessor; and that changes will occur in the trophic balance (fishing down marine food webs).

"Sustainable management of fisheries cannot be achieved without an acceptance that the long-term goals of fisheries management are the same as those of environmental conservation" —Daniel Pauly and Dave Preikshot,

Global wild fisheries are believed to have peaked and begun a decline, with valuable habitats, such as estuaries and coral reefs, in critical condition. Current aquaculture or farming of piscivorous fish, such as salmon, does not solve the problem because farmed piscivores are fed products from wild fish, such as forage fish. Salmon farming also has major negative impacts on wild salmon. Fish that occupy the highertrophic levels are less efficient sources of food energy.

Fishery ecosystems are an important subset of the wider marine environment. This article documents the views of fisheries scientists and marine conservationists about innovative approaches towards sustainable fisheries.

History

In the end, we will conserve only what we love; we will love only what we understand; and we will understand only what we are taught

—Senegalese conservationist Baba Dioum,

In his 1883 inaugural address to the International Fisheries Exhibition in London, Thomas Huxley asserted that overfishing or "permanent exhaustion" was scientifically impossible, and stated that probably "all the great sea fisheries are inexhaustible". In reality, by 1883 marine fisheries were already collapsing. The United States Fish Commission was established 12 years earlier for the purpose of finding why fisheries in New England were declining. At the time of Huxley's address, the Atlantic halibut fishery had already collapsed (and has never recovered).

Traditional Management of Fisheries

Traditionally, fisheries management and the science underpinning it was distorted by its "narrow focus on target populations and the corresponding failure to account for ecosystem effects leading to declines of species abundance and diversity" and by perceiving the fishing industry as "the sole legitimate user, in effect the owner, of marine living resources." Historically, stock assessment scientists usually worked in government laboratories and considered their work to be providing services to the fishing industry. These scientists dismissed conservation issues and distanced themselves from the scientists and the science that raised the issues. This happened even as commercial fish stocks deteriorated, and even though many governments were signatories to binding conservation agreements.

Defining Sustainability

The notion of sustainable development is sometimes regarded as an unattainable, even illogical notion because development inevitably depletes and degrades the environment. Ray Hilborn, of the University of Washington, distinguishes three ways of defining a sustainable fishery.

- *Long term constant yield* is the idea that undisturbed nature establishes a steady state that changes little over time. Properly done, fishing at up to maximum sustainable yield allows nature to adjust to a new steady state, without compromising future harvests. However, this view is naive, because constancy is not an attribute of marine ecosystems, which dooms this approach. Stock abundance fluctuates naturally, changing the potential yield over short and long term periods.
- *Preserving intergenerational equity* acknowledges natural fluctuations and regards as unsustainable only practices which damage the genetic structure destroy habitat, or deplete stock levels to the point where rebuilding requires more than a single generation. Providing rebuilding takes only one generation, overfishing may be economically foolish, but it is not unsustainable. This definition is widely accepted.
- *Maintaining a biological, social and economic system* considers the health of the human ecosystem as well as the marine ecosystem. A fishery which rotates among multiple species can deplete individual stocks and still be sustainable so long as the ecosystem retains its intrinsic integrity. Such a definition might consider as sustainable fishing practices that lead to the reduction and possible extinction of some species.

Social Sustainability

Fisheries and aquaculture are, directly or indirectly, a source of livelihood for over 500 million people, mostly in developing countries. While biodiversity is important, people need food security.

> *Social sustainability can conflict with biodiversity. A fishery is socially sustainable if the fishery ecosystem maintains the ability to deliver products the society can use. Major species shifts within the ecosystem could be acceptable as long as the flow of such products continues. Humans have been operating such regimes for thousands of years, transforming many ecosystems, depleting or driving to extinction many species.*
>
> —*Ray Hilborn,*

According to Hilborn, the "loss of some species, and indeed transformation of the ecosystem is not incompatible with sustainable harvests." For example, in recent years, barndoor skates have been caught as bycatch in the western Atlantic. Their numbers have severely declined and they will probably go extinct if these catch rates continue. Even if the barndoor skate goes extinct, changing the ecosystem, there could still be sustainable fishing of other commercial species.

Reconciling Fisheries with Conservation

At the Fourth World Fisheries Congress in 2004, Daniel Pauly asked, "How can fisheries science and conservation biology achieve a reconciliation?", then answered his own question, "By accepting each other's essentials: that fishing should remain a viable occupation; and that aquatic ecosystems and their biodiversity are allowed to persist."

Obstacles

Overfishing

Overfishing can be sustainable. According to Hilborn, overfishing can be "a misallocation of societies' resources", but it does not necessarily threaten conservation or sustainability".

Overfishing is traditionally defined as harvesting so many fish that the yield is less than it would be if fishing were reduced. For example, Pacific salmonare usually managed by trying to determine how many spawning salmon, called the "escapement", are needed each generation to produce the maximum harvestable surplus.

The optimum escapement is that needed to reach that surplus. If the escapement is half the optimum, then normal fishing looks like overfishing.

But this is still sustainable fishing, which could continue indefinitely at its reduced stock numbers and yield. There is a wide range of escapement sizes that present no threat that the stock might collapse or that the stock structure might erode.

On the other hand, overfishing can precede severe stock depletion and fishery collapse. Hilborn points out that continuing to exert fishing pressure while production decreases, stock collapses and the fishery fails, is largely "the product of institutional failure."

Today over 70% of fish species are either fully exploited, overexploited, depleted, or recovering from depletion. If overfishing does not decrease, it is predicted that stocks of all species currently commercially fished for will collapse by 2048."

A Hubbert linearisation (Hubbert curve) has been applied to the whaling industry, as well as charting the price of caviar, which depends on sturgeon stocks. Another example is North Sea cod. Comparing fisheries and mineral extraction tells us that human pressure on the environment is causing a wide range of resources to go through a Hubbert depletion cycle.

Habitat Modification

Nearly all the world's continental shelves, and large areas of continental slopes, underwater ridges, and seamounts, have had heavy bottom trawls and dredges repeatedly dragged over their surfaces. For fifty years, governments and organisations, such as the Asian Development Bank, have encouraged the fishing industry to develop trawler fleets.

Repeated bottom trawling and dredging literally flattens diversity in the benthic habitat, radically changing the associated communities.

Changing the Ecosystem Balance

Since 1950, 90 percent of 25 species of big predator fish have gone.

Climate Change

Rising ocean temperatures and ocean acidification are radically altering aquatic ecosystems. Climate change is modifying fish distribution and the productivity of marine and freshwater species.

This reduces sustainable catch levels across many habitats, puts pressure on resources needed for aquaculture, on the communities that depend on fisheries, and on the oceans' ability to capture and store carbon (biological pump).

Sea level rise puts coastal fishing communities at risk, while changing rainfall patterns and water use impact on inland (freshwater) fisheries and aquaculture.

Ocean Pollution

A recent survey of global ocean health concluded that all parts of the ocean have been impacted by human development and that 41 percent has been fouled with human polluted runoff, overfishing, and other abuses.

Pollution is not easy to fix, because pollution sources are so dispersed, and are built into the economic systems we depend on.

The United Nations Environment Programme (UNEP) mapped the impacts of stressors such as climate change, pollution, exotic species, and over-exploitation of resources on the oceans. The report shows at least 75 percent of the world's key fishing grounds may be affected.

Diseases and Toxins

Large predator fish contain significant amounts of mercury, a neurotoxin which can affect fetal development, memory, mental focus, and produce tremors.

Irrigation

Lakes are dependent on the inflow of water from its drainage basin. In some areas, aggressive irrigation has caused this inflow to decrease significantly, causing water depletion and a shrinking of the lake.

The most notable example is the Aral Sea, formerly among the four largest lakes in the world, now only a tenth of its former surface area.

Remediation

Fisheries Management

Fisheries management draws on fisheries science to enable sustainable exploitation. Modern fisheries management is often defined as mandatory rules based on concrete objectives and a mix of management techniques, enforced by a monitoring control and surveillance system.

- *Ideas and rules:* Economist Paul Romer believes sustainable growth is possible providing the right ideas (technology) are combined with the right rules, rather than simply hectoring fishers. There has been no lack of innovative ideas about how to harvest fish. He characterises failures as primarily failures to apply appropriate rules.
- *Fishing subsidies:* Government subsidies influence many of the world fisheries. Operating cost subsidies allow European and Asian fishing fleets to fish in distant waters, such as West Africa. Many experts reject fishing subsidies and advocate restructuring incentives globally to help struggling fisheries recover.
- *Economics:* Another focus of conservationists is on curtailing detrimental human activities by improving fisheries' market structure with techniques such as salable fishing quotas, like those set up by the Northwest Atlantic Fisheries Organisation, or laws such as those listed below.
- *Sustainable fisheries certification:* A promising direction is the independent certification programs for sustainable fisheries conducted by organisations such as the Marine Stewardship Council and Friend of the Sea. These programs work at raising consumer awareness and insight into the nature of their seafood purchases.

Ecosystem-Based Fisheries

> *We propose that rebuilding ecosystems, and not sustainability per se, should be the goal of fishery management. Sustainability is a deceptive goal because human harvesting of fish leads to a progressive simplification of ecosystems in favour of smaller, high turnover, lower trophic level fish species that are adapted to withstand disturbance and habitat degradation.*
>
> *—Tony Pitcher and Daniel Pauly,*

According to marine ecologist Chris Frid, the fishing industry points to marine pollution and global warming as the causes of recent, unprecedented declines in fish populations. Frid counters that overfishing has also altered the way the ecosystem works. "Everybody would like to see the rebuilding of fish stocks and this can only be achieved if we understand all of the influences, human and natural, on fish dynamics." He adds: "fish communities can be altered in a number of ways, for example they can decrease if particular-sized individuals of a species are targeted, as this affects predator and prey dynamics. Fishing, however, is not the sole cause of changes to marine life—pollution is another example....No one factor operates in isolation and components of the ecosystem respond differently to each individual factor."

The traditional approach to fisheries science and management has been to focus on a single species. This can be contrasted with theecosystem-based approach. Ecosystem-based fishery concepts have been implemented in some regions. In a 2007 effort to "stimulate much needed discussion" and "clarify the essential components" of ecosystem-based fisheries science, a group of scientists offered the following ten commandments for ecosystem-based fisheries scientists:

- Keep a perspective that is holistic, risk-adverse and adaptive.
- Maintain an "old growth" structure in fish populations, since big, old and fat female fish have been shown to be the best spawners, but are also susceptible to overfishing.
- Characterise and maintain the natural spatial structure of fish stocks, so that management boundaries match natural boundaries in the sea.
- Monitor and maintain seafloor habitats to make sure fish have food and shelter.
- Maintain resilient ecosystems that are able to withstand occasional shocks.
- Identify and maintain critical food-web connections, including predators and forage species.
- Adapt to ecosystem changes through time, both short-term and on longer cycles of decades or centuries, including global climate change.
- Account for evolutionary changes caused by fishing, which tends to remove large, older fish.
- Include the actions of humans and their social and economic systems in all ecological equations.

Marine Protected Areas

Strategies and techniques for marine conservation tend to combine theoretical disciplines, such as population biology, with practical conservation strategies, such as setting up protected areas, as with Marine Protected Areas (MPAs) or Voluntary Marine Conservation Areas. Each nation defines MPAs independently, but they commonly involve increased protection for the area from fishing and other threats. Marine life is not evenly distributed in the oceans. Most of the really valuable ecosystems are in relatively shallow coastal waters, above or near the continental shelf, where the sunlit waters are often nutrient rich fromland runoff or upwellings at the continental edge, allowing photosynthesis, which energises the lowest trophic levels. In the 1970s, for reasons more to do with oil drilling than with fishing, the U.S. extended its jurisdiction, then 12 miles from the coast, to 200 miles. This made huge shelf areas part of its territory. Other nations followed, extending national control to what became known as the exclusive economic zone (EEZ). This move has had many implications for fisheries conservation, since it means that most of the most productive maritime ecosystems are now under national jurisdictions, opening possibilities for protecting these ecosystems by passing appropriate laws.

Daniel Pauly characterises marine protected areas as "a conservation tool of revolutionary importance that is being incorporated into the fisheries mainstream." The Pew Charitable Trusts have funded various initiatives aimed at encouraging the development of MPAs and other ocean conservation measures.

Fish Farming

To what extent can farmed fish be part of the answer? Farmed salmon eat three pounds of wild fish to produce one pound of salmon.

Laws and Treaties

International laws and treaties related to marine conservation include the 1966 Convention on Fishing and Conservation of Living Resources of the High Seas. United States laws related to marine conservation include the 1972 Marine Mammal Protection Act, as well as the 1972 Marine Protection, Research and Sanctuaries Act which established the National Marine Sanctuaries program. Magnuson-Stevens Fishery Conservation and Management Act.

Awareness Campaigns

Various organisations promote sustainable fishing strategies, educate the public and stakeholders, and lobby for conservation law and policy. The list includes the Marine Conservation Biology Institute and Blue Frontier Campaign in the U.S., The U.K.'s Frontier (the Society for Environmental Exploration) and Marine Conservation Society, Australian Marine Conservation Society, International Council for the Exploration of the Sea(ICES), Langkawi Declaration, Oceana, PROFISH, and the Sea Around Us Project, International Collective in Support of Fishworkers, World Forum of Fish Harvesters and Fish Workers, Frozen at Sea Fillets Association and CEDO.

The United Nations Millennium Development Goals include, as goal #7: target 2, the intention to "reduce biodiversity loss, achieving, by 2010, a significant reduction in the rate of loss", including improving fisheries management to reduce depletion of fish stocks.

Some organisations certify fishing industry players for sustainable or good practices, such as the Marine Stewardship Council and Friend of the Sea.

Other organisations offer advice to members of the public who eat with an eye to sustainability. According to the marine conservation biologist Callum Roberts, four criteria apply when choosing seafood:

- Is the species in trouble in the wild where the animals were caught?
- Does fishing for the species damage ocean habitats?
- Is there a large amount of bycatch taken with the target species?
- Does the fishery have a problem with discards—generally, undersized animals caught and thrown away because their market value is low?

The following organisations have download links for wallet-sized cards, listing good and bad choices:

- Monterey Bay Aquarium Seafood Watch, USA
- Blue Ocean Institute, USA
- Marine Conservation Society, UK
- Australian Marine Conservation Society
- The Southern African Sustainable Seafood Initiative.

Data Issues

Data Quality

One of the major impediments to the rational control of marine resources is inadequate data. According to fisheries scientist Milo Adkison (2007), the primary limitation in fisheries management decisions is poor data. Fisheries management decisions are often based on population models, but the models need quality data to be accurate. Scientists and fishery managers would be better served with simpler models and improved data.

Unreported Fishing

Estimates of illegal catch losses range between $10bn and $23bn annually, representing between 11 and 26 million tonnes.

Shifting Baselines

Shifting baselines is a term which describes the way significant changes to a system are measured against previous baselines, which themselves may represent significant changes from the original state of the system. The term was first used by the fisheries scientist Daniel Pauly in his paper "Anecdotes and the shifting baseline syndrome of fisheries". Pauly developed the term in reference to fisheries management where fisheries scientists sometimes fail to identify the correct "baseline" population size (e.g. how abundant a fish species population was *before* human exploitation) and thus work with a shifted baseline. He describes the way that radically depleted fisheries were evaluated by experts who used the state of the fishery at the start of their careers as the baseline, rather than the fishery in its untouched state. Areas that swarmed with a particular species hundreds of years ago, may have experienced long term decline, but it is the level of decades previously that is considered the appropriate reference point for current populations. In this way large declines in ecosystems or species over long periods of time were, and are, masked. There is a loss of perception of change that occurs when each generation redefines what is "natural".

Looting the Seas

Looting the seas is the name given by the International Consortium of Investigative Journalists to a series of journalistic investigations into areas directly affecting the sustainability of fisheries. So far they have investigated three areas involving fraud, negligence and overfishing:

- The black market in bluefin tuna
- Subsidies propping up the Spanish fishing industry
- Overfishing of the southern jack mackerel.

Aquaculture in South Africa

Aquaculture is about the emerging aquaculture industry in South Africa.

Definition of Aquaculture

Aquaculture is defined as the "controlled cultivation (farming/production) of a variety of fish, crustaceans, molluscs, and aquatic plants in marine and freshwater mainly for human consumption, industrial use and recreational purposes. In South Africa it consists mainly of freshwater species such as crocodiles, trout, catfish, tilapia and ornamental fish as well as marine species such as abalone, prawns, oysters and mussels."

Overview

According to the Department of Agriculture, Forestry and Fisheries, "Aquaculture in South Africa is divided into freshwater aquaculture and marine aquaculture. Freshwater fish culture is severely limited by the supply of suitable water. The most important areas for the production of fresh water species are the Limpopo, Mpumalanga Lowveld and Northern Kwazulu–Natal. Trout is farmed along the high mountain in Lydenburg area, Kwazulu-Natal Drakensberg and the Western Cape. Other freshwater species cultivated on a small scale include catfish, freshwater crayfish and tilapia species. Marine aquaculture is a fast developing sector, with a focus on mussels, oysters, abalone, seaweeds and prawns. Of these, mussel farming is the best established. Abalone culture is now well established, centred in the Hermanus area on the Cape south coast. There is also an experimental offshore farm (cage culture) off Gansbaai for salmon."

Export data from the Department of Agriculture, Forestry and Fisheries indicate that the South African industry is dominated by the Western Cape province, which accounts for more than 80% of all South African aquaculture produce, followed by the Eastern Cape at a distant 12.75%.

The Department of Trade and Industry states in its sector diagnostic study that national aquaculture production data (2003–2006) in 2003 was 3,485 tons and in 2006 was 3,564 tons which had a value of R210. Abalone accounted for the biggest increase with production increasing by 61% from 515 tons in 2003, to 833 tons in 2006. Declines in production were experienced by the following sub-sectors: oyster, mussel and trout with reductions in production of 19.2%, 39.5% and 18.4% respectively.

Position of the South African Industry in Relation to the Global Industry

The industry in South Africa is in its infancy, with production volumes being very low, even compared to its continental peers where Egypt and Nigeria lead the pack.

Government Policy and Support Towards the Sector

Aquaculture has been identified as a critical industry, due to the popularity of its produce and the declining yields world-wide. In South Africa, Aquaculture has been identified as part of the key industries for promotion in line with the country's Industrial Policy Action Plan II (IPAP II). The Department of Agriculture, Forestry and Fisheries is also lending its support to the sector and encouraging collaboration with stakeholders across the sector.

Aquaculture in New Zealand

Aquaculture started in New Zealand in the late 1960s. It is dominated by mussels, oysters and salmon. In 2007, aquaculture generated about NZ$360 million in sales on an area of 7,700 hectares. $240 million was earned in exports.

In 2006, the aquaculture industry in New Zealand developed a strategy aimed at achieving a sustainable annual billion NZ dollar business by 2025. In 2007, the government reacted by offering more support to the growing industry.

Aquaculture is the general term given to the cultivation of any fresh or salt water plant or animal. It takes place in New Zealand in coastal marine areas (mariculture) and in inland tanks or enclosures.

Aquaculture in New Zealand currently (2008) occupies 14,188 ha. Of that area, 7,713 ha is in established growing areas and is owned by the aquaculture industry, 4,010 ha is used to enhance the wild scallop fishery and belongs to the Challenger Scallop Enhancement Company, and 2,465 ha is an exposed site six kilometres offshore from Napier where trials are being undertaken by a private company to test the site's economic viability.

In 2005 the aquaculture industry provided direct employment for about 2,500 full time equivalents, mostly in the processing sector. A similar amount of indirect employment resulted from flow-on effects. The aquaculture industry is important for some coastal areas around New Zealand where there is limited employment. This applies particularly to some Mâoricommunities with traditional links to coastal settlements.

Marine aquaculture, mariculture, occurs in the sea, generally in sheltered bays along the coast. In New Zealand, about 70 percent of marine aquaculture occurs in the top of the South Island. In the North Island, the Firth of Thames is productive.

Cultured Species

There are three main species in the New Zealand aquaculture industry: the green-lipped mussel, the Pacific oyster and king salmon. In 2006 these three species generated $357 million in sales. Mussel accounted for 63 percent of this value, Pacific oysters 9 percent and king salmon 28 percent.

Over two-thirds of New Zealand's aquaculture product comes from mussels and oysters. These shellfish are cultivated in two distinct stages; first spat needs to be collected, then the spat is grown in a grow-out facility.

- Spat, also called seed, is the free-swimming larval stage of a shellfish. Spat is cultured in hatcheries, and can be grown in tanks on land. Hatcheries can also be associated with research facilities where spat can be selectively bred to specifications, as broodstock.
- A grow-out facility is the place where the spat are raised to market size, usually in enclosures anchored in coastal waters.

Mussels

Until the early 1960s, mussels were harvested by hand from intertidal rocks. Dredging was then introduced, and within a few years the mussel beds in Tasman Bay and the Hauraki Gulf were dredged clean. In the late 1960s, following this collapse, the aquaculture of the New Zealand mussel began. The endemic green lipped mussel was used to trial growing mussel spat (young mussels) on ropes suspended from rafts.

The Hauraki Gulf and the Marlborough Sounds provided sheltered environments, with clean water rich in plankton. The cultured mussels were ready for harvest after 12 to 18 months, and first went on sale in 1971.

More growers entered the industry. The labour-intensive raft method was replaced with a modified Japanese longline system. Biodegradable stockings were packed with spat and tied to parallel rows of looped ropes, supported by buoys. Young mussels grow through the stockings, anchoring themselves to the ropes with their strong byssal threads (beards). The farms are usually located in sheltered or semi-sheltered areas where there is sufficient depth of water at low tide to keep the longline droppers off the bottom. Recent research has been investigating offshore mussel farming in exposed areas several kilometres from shore, such as farms offshore from Napier and Opotiki.

Figure: Green lipped mussel

Initially the ropes were allowed to reseed naturally, after harvest, from the spat already present in coastal waters. However, this method was unreliable. In 1974 a marine scientist discovered mussel spat encrusted on drift kelp on Ninety Mile Beach. Locals collected the seaweed and air freighted it to mussel farmers. Kaitaia spat, as it became known, is now the prime source of seed mussels. There are some experimental hatcheries.Improved techniques have led to rapid production increases, and bulk handling methods have been introduced to meet growing demand for export to more than 60 countries. By 2006 there were over 900 mussel farms in New Zealand covering about 6500 hectares, and worth about $224 million in annual sales. About $180 million were exports, usually sold as frozen mussels in half shells, patented with the trade name NZ Greenshell Mussels.

Oysters

There are two types of wild oysters in New Zealand, Bluff oysters and rock oysters. Both have been commercially harvested since the mid-19th century. Bluff oysters have never been cultivated, but various attempts were made to cultivate the rock oyster.

Rock oysters are found naturally in the intertidal zone in the north of the North Island, and were subject to early cultivation experiments. During the 1960s, commercial farmers grew rock oysters on sticks coated with cement, and laid in racks in the lower intertidal regions of harbours and inlets around the northern North Island.

Then in 1970 another oyster started outgrowing the native rock oyster. This newcomer was the Pacific oyster, which had probably been introduced into New Zealand waters in the 1950s from a Japanese vessel hull or in their ballast water. At first, farmers tried to remove it from their collecting sticks, but year by year Pacific oyster spat increased, out-competing the native rock oyster.

Eventually commercial growers began to cultivate the Pacific oyster, and by the mid-1970s, it had become the main farm-raised oyster. Pacific oysters have well-established international markets, grow three times faster than native rock oysters, reach a larger size, have several spawnings each year and produce more consistent quantities of spat.

In 1977 Pacific oysters appeared in the Marlborough Sounds, and farming began there in the 1990s. Instead of using the North Island method of cultivating oysters on racks, Marlborough farmers used hanging longlines, a system developed in Australia.

By 2006 there were over 230 oyster farms in New Zealand using over 900 hectares, and worth about $32 million in annual sales. About $18 million were exports.

Hatchery reared seed is not suitable for the rack systems which are still used by much of the industry. Pacific oysters on these racks are cultured from wild spat, most of which is gathered in the Kaipara Harbour. However, other systems are increasingly being used, including hanging longlines and plastic trays. The Cawthron Institute is the main provider of hatchery spat, and can selectively breed spat to specifications.

Salmon

Around 1900, different salmon were introduced as sport fish. Only the king salmon (Chinook) adapted to the environment. For decades, the development of salmon and trout aquaculture in New Zealand was opposed by recreational fishers, on the grounds that disease would spread from fish farms into recreational fisheries, and that wild fish would be poached if they could be sold. In 1973 the government compromised by making trout farms illegal, but salmon farms legal. New Zealand is probably the only country in the world where trout farming is illegal, despite commercial interest from companies already farming salmon.

In 1976, the first salmon farm was established at Pupu Springs, Tasman. Salmon were raised in fresh water, growing to 25 centimetres over two years. The venture was originally aimed at ocean ranching, where juvenile salmon would be released into the sea with the expectation that some would return as adults. But few did return, so the Pupu Springs facility was converted to a hatchery, supplying stock to sea farms.

In 1983, the first sea-cage salmon farm was established in Big Glory Bay, Stewart Island. It was followed by farms in the Marlborough Sounds and at Akaroa, Banks Peninsula. These areas accounted for over 90 percent of the 8,500 tonnes of salmon produced in 2001. Today, New Zealand accounts for over half of the world production of king salmon (7,400 tonnes in 2005).

Farming in the sea (mariculture) for king salmon is sometimes called sea-cage ranching. Sea-cage ranching takes place in large floating net cages, about 25 metres across and 15 metres deep, moored to the sea floor in clean, fast-flowing coastal waters. Smolt (young fish) from freshwater hatcheries are transferred to cages containing several thousand salmon, and remain there for the rest of their life. They are fed fishmeal pellets high in protein and oil. Most of this fishmeal is imported from Australia. The salmon are harvested when they are about two years old, weighing 2.5 to 4 kilograms. Sea cages are located in the Marlborough Sounds, Akaroa Harbour and Stewart Island.

Farming in freshwater for king salmon uses net cages placed in rivers, using techniques similar to those used for sea-farmed salmon. Freshwater raceways are located in several Canterbury rivers such as the Cluthaand Waimakariri Rivers. In the 1990s, a unique form of freshwater salmon farming was developed in hydroelectric canals in the Mackenzie Basin. Young salmon are enclosed in pens in the Ohau and Tekapo canals. The Tekapo site, fed by fast cold waters from the Southern Alps, is the highest salmon farm in the world, 677 metres above sea level.

Before they are killed, cage salmon are anaesthetised with a herbal extract. They are then spiked in the brain. The heart beats for a time as the animal is bled from its sliced gills. Relaxing the salmon like this when it is killed produces firm, long-keeping flesh. Lack of disease in wild populations and low stocking densities used in the cages means that New Zealand salmon farmers do not use antibiotics and chemicals that are often needed elsewhere.

The New Zealand industry has grown into the largest producers of farmed king salmon in the world, accounting for about half of world production. The New Zealand King Salmon Company, dominates the production of king salmon in New Zealand. The company has its own selective breeding programmes integrated with quality control and production. Other salmon producers rely on stock from hatcheries where selective breeding is less well developed.

Other Species

- The culture of ornamental cold water species, such as goldfish, are valued at about $18 million.
- A small-scale freshwater prawn farm was established in 1991 at Wairakei, near Taupo, producing tropical giant river prawns. Heat from a geothermal source is used to heat water in prawn-rearing ponds.

Other species which have potential, but are currently small-scale or are still in research or pre-commercial stages, are discussed under Prospects below.

Scallop Enhancement

Enhancement is the name given to techniques designed to boost the natural recruitment or survival of young animals or seaweed in the wild. In New Zealand, scallop enhancement has worked well in Tasman Bay and Golden Bay.

The New Zealand scallop is a large fan-shaped shellfish, flat on one side and convex on the other. It lives on the bottom of coastal waters, 30 metres or more deep. Scallop spat-collecting bags are suspended during summer in coastal areas with high natural scallop settlement. The scallop larvae settle out of the plankton onto the fine feathery surface of the plastic mesh bags. The larvae are allowed to grow to a suitable size and are then released onto known natural scallop beds at densities of about six per square metre of sea floor. There, they are later harvested on a rotational basis by dredges. This technique has resulted in a marked increase and stabilising of the available annual catch. The Tasman scallop fishery, near collapse in the 1980s, recovered with re-seeding to a level where 747 tonnes were harvested in 2004.

Legislation and Administration

Marine farmers usually look for sheltered and unpolluted waters rich in nutrients. Often these areas are also desirable for other purposes. In the late 1990s, demand for coastal aquaculture space upsurged, increasing fivefold. Aquaculture consents developed haphazardly, with regional councils unsure about how marine farms might impact coastal environments. By 2001, some councils were inundated with marine farm applications, and were operating with inadequate guidelines for sustainably managing the coast. As the Ministry for the Environment put it: "Attempts to minimise local or cumulative environmental effects resulted in bottlenecks, delays and high costs in processing applications for new marine farms, local moratoria, submitter fatigue and poor environmental outcomes. Marine farmers, local communities, and the government wanted change."

In 2002, the government stopped issuing consents for more new marine farms while they reformed the legislation. The consents had operated under a system overseen by both the Ministry of Fisheries and the regional councils. The reforms aimed to streamline these applications for both freshwater and marine farms. Industry farmers objected to the moratorium, on the grounds that delaying expansion and diversification could not be in the interest of the industry. Mâori groups considered they were especially affected since they were the main applicants for coastal farms.

This took three years, and in early 2005, Parliament passed the Aquaculture Reform Act 2004, which introduced the new legislation. The act amends five existing acts to cope with the new environmental demands, and creates two new acts, the Mâori Commercial Aquaculture Claims Settlement Act 2004 and the Aquaculture Reform (Repeals and Transitional Provisions) Act 2004. The legislation and administration of aquaculture in New Zealand is complex for such a small industry. A more comprehensive overview can be found here.

Aquaculture is administered in New Zealand through labyrinth bureaucracies, with consequent diluted responsibilities. No single ministerial portfolio or government agency is responsible. As an example, in 2007 the government released a strategy on aquaculture. This strategy was endorsed by six government ministers with the following portfolios: fisheries, environment, conservation, local government, Mâori affairs, industry and regional development. Further, there were five government departments directly involved in the preparation of the strategy. As another example, the access to marine and freshwater aquaculture sites are under the control of 17 regional local government agencies with yet more oversight by various central government agencies.

Despite many further consultations and incentives, no new aquaculture space was created under the new legislation for another four years. This coincided with a change in government at the end of 2008, which announced that the aquaculture reforms are to be overhauled.

Training and Research

In recent years, skill levels in the New Zealand aquaculture industry has considerably improved. This has been largely due to Seafood Industry Training Organisation (SITO), an integral part of the seafood industry. SITO have developed tailored aquaculture training programmes based on their prior experience with industry-based training for wild fisheries. They now offer nationally recognised training programmes based on the needs of companies involved in aquaculture.

At the tertiary level, the Auckland University of Technology offers an undergraduate degree in aquaculture. Other tertiary training centres offering aquaculture courses include the Bay of Plenty Polytechnic, the Nelson Marlborough Institute of Technology, and the Mahurangi Technical Institute.

Government funding for aquaculture research is about two percent of the annual sales of the industry. These funds are mostly delivered through a competitive bidding process, organised and controlled by the Foundation for Research, Science and Technology.

- The principal aquaculture research group is the National Institute of Water and Atmospheric Research (NIWA). NIWA is structured as a profit-making private company, though it is owned by the government. It operates three aquaculture research facilities; Bream Bay Aquaculture Park, Mahanga Bay Aquaculture Research Facility and Silverstream Hatchery. The Bream Bay Aquaculture Park includes other private aquaculture companies organised as an industrial-technology park. NIWA produces yellowtail kingfish spat, seed abalone and salmon smolts, which it sells to on-growers.
- The Cawthron Institute is a non-profit organisation which does regional research around Nelson. It operates a nearby saltwater research facility called the Cawthron Aquaculture Park.
- The tertiary education sector undertakes a small amount of aquaculture research. In 2007 an aquaculture centre was opened at Mahurangi Technical Institute in Warkworth. Scientists at the institute are aiming to breed short-fin eels within two years with a goal of producing commercial quantities of eels in captivity, which would be a world first.

Role of Mâori

In pre-European times, the indigenous Mâori of New Zealand undertook rudimentary aquaculture activities, such as placing suitable rocks into the intertidal settlement zones of oyster larvae. They were also thought to have transplanted abalone and other shellfish between different areas.

Mâori currently have a significant presence in the New Zealand aquaculture industry, and this is likely to increase over time as the requirements to allocate aquaculture space through the Mâori Commercial Aquaculture Claims Settlement Act 2004 are met. However, inappropriate aquaculture locations and unsustainable practices have the potential to compromise values and resources important to coastal whanau, hapu and iwi. In 2008, a settlement of $97 million was made to Mâori for Crown obligations for aquaculture space that was approved between 1992 and 2004. NIWA operates a Mâori research and development unit, Te Kûwaha o Taihoro Nukurangi. The unit has a team of Mâori scientists who undertake research and provide consultancy services, based particularly aroundiwi with environmental and commercial issues.

Prospects

In 2006, the New Zealand aquaculture industry published *The New Zealand Aquaculture Strategy*, setting itself an annual sales target of one billion NZ dollars by 2025. The strategy sets out ten areas of activity needed to achieve this target.

The New Zealand Aquaculture Council has introduced a levy on aquaculture producers so this strategy can be implemented. In 2007, the New Zealand government responded to this industry initiative by releasing an aquaculture development strategy highlighting existing actions and proposing new initiatives. In addition, the government has offered additional funding around five key objectives, with the main focus on improving the implementation of the new 2004 regulations. At the end of 2008 there was a change in government, which announced that the aquaculture reforms will be overhauled, but reaffirmed the government commitment to the industry billion dollar target.

The New Zealand industry currently relies on low-value filter feeding shellfish (mussels and oysters) which are fast growing and relatively easy to culture. There is potential for the industry to diversify into higher value species such as pâua, kingfish and crayfish. These species need special food supplies and are more expensive to farm, but they command higher prices.

- The native blackfoot pâua, *Haliotis iris*, is a form of abalone. They are large sea snails which survive strong tidal surges by clinging to rocks using their large muscular foot. Wild pâua has been harvested since 1944, usually by skindivers,. Pâua aquaculture started in 1980, but has been slow moving beyond development. They are difficult to grow, and grow more slowly than salmon, mussels or oysters. Their larva and juveniles need to be grown separately. Most pâua farmers get juveniles from hatcheries, and feed them fresh kelp in land-based tanks. In Akaroa Harbour, one farmer grows pâua on plastic barrels tethered to buoys. In 2002, farmers produced five tonnes of pâua meat, worth $400,000.
- Another New Zealand pioneer pâua farmer cultivates blue pearls by placing some grit between the flesh of the pâua and its shell, where it acts as an irritant. The pâua responds by coating the grit with nacre (mother-of-pearl). This develops as a blue pearl.
- For some years there has been research on the best ways of growing the red seaweed, *Gigartina atropurpurea*, in New Zealand. Seaweed spores are grown on three metre strings at NIWA's Mahanga Bay aquaculture research facility, and are then transferred to a mussel farm in the Marlborough Sounds. If successful, a new seaweed growing industry could spread to the mussel farms in the Marlborough Sounds.
- Bluff oysters are harvested from the wild in Foveaux Strait. However, they breed more easily in Northland, and NIWA is examining their aquaculture possibilities.
- Shortfin and longfin eels have been trialled by NIWA. Established worldwide markets in cultured eels are worth over US$1 billion, and a decline in some stocks has opened up opportunities for New Zealand.

- The big-bellied seahorse is a native seahorse. Seahorses are valued aquarium fish. They are also used medicinally, particularly in traditional Chinese medicine. Wild seahorses have been over-harvested worldwide, opening markets to their aquaculture.
- A New Zealand sea sponge, *Mycale hentscheli*, which grows in Pelorus Sound, may hold the key to an anti-cancer drug. A substance produced by the sponges, might be used as a cancer-fighting drug. Victoria University and NIWA are working with Marlborough marine farmers to develop a method for growing the sponge on an existing mussel farm.

Table: Other prospects which are being researched or trialled include

• European perch	• grass carp
• groper	• hâpuka, a native wreckfish
• yellowtail kingfish, a native kingfish	• sea cucumber
• kina, a native sea urchin	• rock lobster
• koura, a native freshwater lobster	• *Spongia manipulatus*, a native bath sponge
• giant kelp	

Timeline

- *Pre-European:* The indigenous Mâori undertake rudimentary aquaculture activities, such as placing suitable rocks into the intertidal settlement zones of oyster larvae. They are also thought to have transplanted abalone and other shellfish between different areas.
- *Early 20th century:* Salmon species are introduced to New Zealand as sport fish, but only the Chinook, or king salmon adapts to the environment.
- *1950s:* the Pacific oyster is introduced, possibly from a Japanese vessel hull or in their ballast water.
- *Early 1960s:* Dredge fisheries start operating in the north of the South Island and around the Hauraki Gulf. Within a few years they dredge these areas bare.
- *Late 1960s:* As a response to the collapse of the dredge fisheries, the aquaculture of New Zealand mussels begins.
- *1970s:* Farming of king salmon begins.
- *Late 1990s:* The aquaculture industry goes through a boom period, and demand for coastal space increases fivefold.
- *2002:* The government, in some disarray, imposes a moratorium on new marine farms while they attempt to develop better legislation aimed at dealing with the environmental demands of aquaculture and streamlining applications for marine and freshwater farms.

- *2005:* Parliament passes the Aquaculture Reform Act 2004, amending five existing acts so they can better cope with the environmental demands of aquaculture, and creating two new acts. However, the reform fails to streamline applications, and no further allocation of aquaculture space occurs over the next four years.
- *2006:* The New Zealand aquaculture industry publishes *The New Zealand Aquaculture Strategy*, setting itself an annual sales target of one billion NZ dollars by 2025.
- *2007:* The New Zealand government responds to the industry initiative by releasing an aquaculture development strategy highlighting existing actions and proposing new initiatives including funding incentives, mainly aimed at trying to action its reform legislation.
- *2008:* A settlement of $97 million is made to Mâori for Crown obligations for aquaculture space that was approved between 1992 and 2004.
- *2008:* The government changes and announces that the aquaculture reforms will be overhauled. It reaffirms the government commitment to the industry billion dollar target.

6

Aquaculture of Coral

Coral aquaculture, also known as coral farming or coral gardening, is the cultivation of corals for commercial purposes or coral reef restoration. Aquaculture is showing promise as a potentially effective tool for restoring coral reefs, which have been declining around the world. The process bypasses the early growth stages of corals when they are most at risk of dying. Coral seeds are grown in nurseries then replanted on the reef. Coral is farmed by coral farmers who live locally to the reefs and farm for reef conservation or for income. It is also farmed by scientists for research, by businesses for the supply of the live and ornamental coral trade and by private aquarium hobbyists.

Introduction

Coral reef farming is the practice of extracting segments or larva from live corals on the reef in order to replant them in a nursery, where they can be protected while they grow to adulthood. Once the corals have grown they are transplanted back into the reef, usually onto damaged areas. This technique has been proven to improve the recovery of coral reefs, which are currently suffering from degradation.

The method is commonly referred to as the "gardening method" as it is analogous to horticulture and has been compared to silviculture as a management practice that mimics natural ecosystems. The technique involves treating coral as if it were a plant. Coral seeds are grown in nurseries, then replanted into their natural habitat – in this case, the sea. Coral reef farming is predominantly practiced for three reasons, either for conservation, to supply aquariums and zoos for public exhibits or to supply the home aquarium hobby industry. Some areas where coral is farmed in situ for conservation reasons are the Philippines, Solomon Islands, Palau, Fiji, Marshall Islands and Japan. Coral farming ex situ occurs more frequently in public aquariums in North America and Europe.

Conservation

The decline of reefs, globally has been measured as 1% annually before 1997, increasing to 2% over the period between 1997 and 2003. Adding to this, the reefs of world are further affected by severe weather events, such as cyclones, from predation from crown of thorns starfish and from competition for habitat with other foundation species such as seaweeds. Seaweeds can take over the coral's usual habitat when fishing stocks are too low and the herbivorous fish do not keep the seaweed at bay by eating it. Adding to this, there is also the increased frequency of coral bleaching events, including "the severe bleaching event" of 1997-1998. It is reported that the increased water temperatures and ocean acidification have correlated with an increase in bleaching events.

These natural stressors to the coral reef are further aggravated by the human impact on coral reefs. Anthropogenic stressors such as coastal development, dynamite fishing, cyanide fishing, overexploitation of resources and marine pollution, have left 58% percent of the world's reefs under threat. An example is the exploitation of mushroom coralin Indonesia which is harvested for supply of the jewellery and curios trades. The harvesting of live reef organisms, including coral, is increasing around the world, and there is concern about the damage occurring to coral reefs as a result of the marine aquarium trade. Coral is often overharvested from the reef to supply this growing demand. Overharvesting is weakening the ability of reefs to replenish after other harmful events.

Coral aquaculture and transplantation have the ability to improve coral cover, biodiversity, and structural heterogeneity of a degraded reef. Success of the technique has been achieved with fire coral, *Pocillopora verrucosa* and *Acropora hemprichi*. The restoration of the reef is a restoration of the natural habitat of organisms associated with the reef, such as reef fishes. Coral is an important foundation species. While it covers less than one percent of the ocean surface, it provides habitat for nearly one third of the all saltwater fish species, as well as ten percent of all fish captured for human consumption. Nursery-grown coral might also promote reef resilience by making contributions to the larval pool. This could have a positive effect on new growth if transplanting of the new coral is made just before a larval release season.

Oceanographer, Baruch Rinkevich coined the term *active restoration* to describe the method of coral reef farming in contrast with what he described as *passive restoration efforts* such as the designation of marine protected areas (MPAs). The active restoration of coral reef farming is so called because it involves human intervention to aid conservation. MPAs or bans on coral extraction are described as passive because there is a lack of human interference such as coral harvesting, fishing and dredging. MPAs are often placed around coral reefs in the hope that if the coral is left alone it will recover.

Commercial Supply for Aquarium Hobbyists

Many people enjoy the creating their own coral display in a home aquarium. In response to this, there are many businesses, such as "ORA", that farm coral in tanks to supply coral to the aquarium hobbyist market. Corbin reports that the 1999 Honolulu, Hawaii Marine Ornamentals Conference concluded with a recommendation to "give highest priority to projects involving the advancement of marine ornamental aquaculture and reef preservation." Conferees pressed the importance of encouraging hobbyists to supply only from coral reef farms to help deter the over-harvesting of corals. Conferees were also reported to have recommended initiatives to encourage consumer understanding that cultured ornamentals are a more sustainable and 'higher value' alternative to wild-caught live reef organisms.

Methods

For Conservation

There are three stages to the farming of coral for coral reef restoration: Collecting asexual or sexual material and growing to in tanks, further growing in submerged sea nurseries and finally re-transplantation into the reef.

Stage One

Coral can reproduce sexually by spawning or asexually by budding polyps. The current mode of sourcing coral seedlings is to obtain asexual material from reef colonies or stray coral fragments by way of harvesting coral branches, fragments or nubbins (pieces of coral pruned from the tips). This fragmentation is by far the most widely practiced method.

Another method currently being pioneered involves collecting the coral spawn from sexual reproduction and growing up the coral right from the zygote stage. This option has many benefits that are beginning to earn it much attention by researchers. Coral colonies on the same reef usually spawn together in one synchronized spawning event. This allows for hundreds of thousands of coral eggs to be collected at one time. This method is known as *spat stocking*. Linden describes a growing apparatus made of Petri disheslined with preconditioned Mailer's paper disks on which the planula of *Stylophora pistillata* are grown. One-month-old survivors were transferred onto plastic pins in a mid water coral nursery, where the trays were covered with fitted plastic nets to prevent predation and detachment. Four months later, more than 89% of the corals had survived. At the Great Barrier Reef Aquarium in Townsville, Australia, large mother colonies of Acropora formosa have collection devices placed above them when spawning is about to take place. Small mature colonies are physically removed from the reef and shifted into a laboratory tank for spawning. They can then be reattached to the reef with cement.

Using this method, the mother colonies are not affected in anyway. It may be possible to cultivate large polyped stony corals, such as *Caryophylliina*, *Euphyllia*, *Heliofungia*,*Lobophyllia* and *Trachyphyllia*, on a large-scale using sexual reproduction, since they do not reproduce by fragmenting into large numbers. This method has also been proved effective on Red Sea soft coral species, *Alcyonarians*: *Clavularia hamra*, *Nephthea sp.*, *Litophyton arboreum*.

Stage Two

The second stage of the coral farming process is when the corals are transported from their indoor tanks, into floating nurseries in the sea. The nurseries are ideally located at a mid-depth. While they will float in the water column, the coral will attached on the structure at a point that is submerged. Some authors recommend about 6 metres depth with a consideration to ensuring the corals will get just the right amount of sunlight. On the nursery, they are affixed to an artificial base to use as a substrate. This is usually made from string, wire, mesh, monofilament line or epoxy. The colonies will remain here from 8 to 24 months to be reared to a fit size for transplantation back to the reef.

Stage Three

When the corals are big enough to be transplanted into the reef, the transplantation stage involves securing to the corals by plastic pegs or masonry anchors or with epoxy glue.

For Commercial or Exhibition Supply

The process involves two steps: First, the larva are harvested from 'ex situ' mature broodstock corals. The coral planula, commonly called *seedlings*, are grown up in tanks in an indoor facility. Secondly, when they are juveniles they are transported to a submerged nursery in the sea. They remain there until adulthood, and when they have grown to a marketable size (about fist-sized) they can be sold.

Social Benefits

Another benefit of coral aquaculture is that it can offer alternative livelihoods to people living near the reefs. This is especially important for communities where income from fishing or harvesting marine organisms have become unsustainable, such as Indonesia. It is possible to use the resources in coral reef farming in a way that is environmentally friendly. Many coral reefs are in locations of low socio-economic status. Coral reef aquaculture has been championed as being a cheap venture to embark on requiring only basic, cheap materials.

History

One of the first serious attempts at propagating coral ex situ occurred at the Noumea Aquarium in 1956. At the time it was common for aquarium hobbyists in Germany to create home "mini-reefs". Commercial coral propagation began in America in the 1960's but the demand in the hobby industry did not really take off until the early 1980s. The trend was attributed to popular articles which appeared in hobby magazines.

Research and Development

Coral aquaculture provides insights into the life histories of a variety of corals. Peterson has shown that early sexual recruits will grow larger when fed the nauplii of brine shrimp. This discovery could shorten the fragile post settlement time in the hatchery.

The Mote Marine Laboratory keeps many broodstock colonies at its Tropical Research Laboratory. The laboratory reports that its colonies are grown from fragments rescued from boat groundings and environmental disturbances. The corals in the broodstock reserve provide fragments for restoration research. Studies are done to determine optimal size, shape and season for restoration.

Fish Processing

Humans have been processing fish since neolithictimes. This 16th century fish stall shows many traditional fish products.

The term fish processing refers to the processes associated with fish and fish products between the time fish are caught or harvested, and the time the final product is delivered to the customer. Although the term refers specifically to fish, in practice it is extended to cover any aquatic organisms harvested for commercial purposes, whether caught in wild fisheries or harvested from aquaculture or fish farming.

Larger fish processing companies often operate their own fishing fleets or farming operations. The products of the fish industry are usually sold to grocery chainsor to intermediaries. Fish are highly perishable. A central concern of fish processing is to prevent fish from deteriorating, and this remains an underlying concern during other processing operations.

Fish processing can be subdivided into fish handling, which is the preliminary processing of raw fish, and the manufacture of fish products. Another natural subdivision is into primary processing involved in the filleting and freezing of fresh fish for onward distribution to fresh fish retail and catering outlets, and the secondary processing that produces chilled, frozen and canned products for the retail and catering trades.

There is evidence humans have been processing fish since the early Holocene. These days, fish processing is undertaken by artisan fishermen, on boardfishing or fish processing vessels, and at fish processing plants.

Fish is a highly perishable food which needs proper handling and preservation if it is to have a long shelf life and retain a desirable quality and nutritional value. The central concern of fish processing is to prevent fish from deteriorating. The most obvious method for preserving the quality of fish is to keep them alive until they are ready for cooking and eating. For thousands of years, China achieved this through the aquaculture of carp. Other methods used to preserve fish and fish products include

- the control of temperature using ice, refrigeration or freezing
- the control of water activity by drying, salting, smoking or freeze-drying
- the physical control of microbial loads through microwave heating or ionising irradiation
- the chemical control of microbial loads by adding acids
- oxygen deprivation, such as vacuum packing.

Usually more than one of these methods is used. When chilled or frozen fish or fish products are transported by road, rail, sea or air, the cold chain must be maintained. This requires insulated containers or transport vehicles and adequate refrigeration. Modern shipping containers can combine refrigeration with a controlled atmosphere.

Fish processing is also concerned with proper waste management and with adding value to fish products. There is an increasing demand for ready to eat fish products, or products that don't need much preparation.

Handling the Catch

When fish are captured or harvested for commercial purposes, they need some preprocessing so they can be delivered to the next part of the marketing chain in a fresh and undamaged condition. This means, for example, that fish caught by a fishing vessel need handling so they can be stored safely until the boat lands the fish on shore. Typical handling processes are

- transferring the catch from the fishing gear (such as a trawl, net or fishing line) to the fishing vessel
- holding the catch before further handling
- sorting and grading
- bleeding, gutting and washing
- chilling
- storing the chilled fish
- unloading, or landing the fish when the fishing vessel returns to port

The number and order in which these operations are undertaken varies with the fish species and the type of fishing gear used to catch it, as well as how large the fishing vessel is and how long it is at sea, and the nature of the market it is supplying. Catch processing operations can be manual or automated. The equipment and procedures in modern industrial fisheries are designed to reduce the rough handling of fish, heavy manual lifting and unsuitable working positions which might result in injuries.

Handling Live Fish

Live Fish Trade: An alternative, and obvious way of keeping fish fresh is to keep them alive until they are delivered to the buyer or ready to be eaten. This is a common practice worldwide. Typically, the fish are placed in a container with clean water, and dead, damaged or sick fish are removed. The water temperature is then lowered and the fish are starved to reduce their metabolic rate. This decreases fouling of water with metabolic products (ammonia, nitrite and carbon dioxide) that become toxic and make it difficult for the fish to extract oxygen.

Fish can be kept alive in floating cages, wells and fish ponds. In aquaculture, holding basins are used where the water is continuously filtered and its temperature and oxygen level are controlled. In China, floating cages are constructed in rivers out of palm woven baskets, while in South America simple fish yards are built in the backwaters of rivers. Live fish can be transported by methods which range from simple artisanal methods where fish are placed in plastic bags with an oxygenated atmosphere, to sophisticated systems which use trucks that filter and recycle the water, and add oxygen and regulate temperature.

Preservation

Preservation techniques are needed to prevent fish spoilage and lengthen shelf life. They are designed to inhibit the activity of spoilage bacteria and the metabolic changes that result in the loss of fish quality. Spoilage bacteria are the specific bacteria that produce the unpleasant odours and flavours associated with spoiled fish. Fish normally host many bacteria that are not spoilage bacteria, and most of the bacteria present on spoiled fish played no role in the spoilage. To flourish, bacteria need the right temperature, sufficient water and oxygen, and surroundings that are not too acidic. Preservation techniques work by interrupting one or more of these needs. Preservation techniques can be classified as follows.

Control of Temperature

If the temperature is decreased, the metabolic activity in the fish from microbial or autolytic processes can be reduced or stopped. This is achieved by refrigeration where the temperature is dropped to about 0 °C, or freezing where the temperature is dropped below -18°C. On fishing vessels, the fish are refrigerated mechanically by circulating cold air or by packing the fish in boxes with ice. Forage fish, which are often caught in large numbers, are usually chilled with refrigerated or chilled seawater. Once chilled or frozen, the fish need further cooling to maintain the low temperature. There are key issues with fish cold store design and management, such as how large and energy efficient they are, and the way they are insulated and palletized.

An effective method of preserving the freshness of fish is to chill with ice by distributing ice uniformly around the fish. It is a safe cooling method that keeps the fish moist and in an easily stored form suitable for transport. It has become widely used since the development of mechanical refrigeration, which makes ice easy and cheap to produce. Ice is produced in various shapes; crushed ice and ice flakes, plates, tubes and blocks are commonly used to cool fish. Particularly effective is slurry ice, made from micro crystals of ice formed and suspended within a solution of water and a freezing point depressant, such as common salt.

A more recent development is pumpable ice technology. Pumpable ice flows like water, and because it is homogeneous, it cools fish faster than fresh water solid ice methods and eliminates freeze burns. It complies with HACCP and ISO food safety and public health standards, and uses less energy than conventional fresh water solid ice technologies.

Control of Water Activity

The water activity, a_w, in a fish is defined as the ratio of the water vapour pressure in the flesh of the fish to the vapour pressure of pure water at the same temperature and pressure. It ranges between 0 and 1, and is a parameter that measures how available the water is in the flesh of the fish. Available water is necessary for the microbial and enzymatic reactions involved in spoilage. There are a number of techniques that have been or are used to tie up the available water or remove it by reducing the a_w. Traditionally, techniques such as drying, salting and smoking have been used, and have been used for thousands of years. These techniques can be very simple, for example, by using solar drying. In more recent times, freeze-drying, water binding humectants, and fully automated equipment with temperature and humidity control have been added. Often a combination of these techniques is used.

Physical Control of Microbial Loads

Heat or ionising irradiation can be used to kill the bacteria that cause decomposition. Heat is applied by cooking, blanching or microwave heating in a manner that pasteurizes or sterilizes fish products. Cooking or pasteurizing does not completely inactivate microorganisms and may need to be followed with refrigeration to preserve fish products and increase their shelf life. Sterilised products are stable at ambient temperatures up to 40°C, but to ensure they remain sterilized they need packaging in metal cans or retortable pouches before the heat treatment.

Chemical Control of Microbial Loads

Biopreservation and Fermented Fish: Microbial growth and proliferation can be inhibited by a technique called biopreservation. Biopreservation is achieved by adding antimicrobials or by increasing the acidity of the fish muscle. Most bacteria stop multiplying when the pH is less than 4.5. Acidity is increased by fermentation, marination or by directly adding acids (acetic, citric, lactic) to fish products.

Lactic acid bacteria produce the antimicrobial nisin which further enhances preservation. Other preservatives include nitrites, sulphites, sorbates, benzoates andessential oils.

Control of the Oxygen Reduction Potential

Spoilage bacteria and lipid oxidation usually need oxygen, so reducing the oxygen around fish can increase shelf life. This is done by controlling or modifying the atmospherearound the fish, or by vacuum packaging. Controlled or modified atmospheres have specific combinations of oxygen, carbon dioxide and nitrogen, and the method is often combined with refrigeration for more effective fish preservation.

Combined Techniques

Hurdle Technology: Two or more of these techniques are often combined. This can improve preservation and reduce unwanted side effects such as the denaturation of nutrients by severe heat treatments. Common combinations are salting/drying, salting/marinating, salting/smoking, drying/smoking, pasteurization/refrigeration and controlled atmosphere/refrigeration. Other process combinations are currently being developed along the multiple hurdle theory.

Automated Processes

"The search for higher productivity and the increase of labour cost has driven the development of computer vision technology, electronic scales and automatic skinning and filleting machines."

Waste Management

Figure: Non-edible fish scrap processing, 1884

Waste produced during fish processing operations can be solid or liquid.

- Solid wastes: include skin, viscera, fish heads and carcasses (fish bones). Solid waste can be recycled in fish meal plants or it can be treated as municipal waste.
- Liquid wastes: include bloodwater and brine from drained storage tanks, and water discharges from washing and cleaning. This waste may need holding temporarily, and should be disposed of without damage to the environment. How liquid waste should be disposed from fish processing operations depends on the content levels in the waste of solid and organic matter, as well as nitrogen and phosphorus content, and oil and grease content. It also depends on an assessment of parameters such acidity levels, temperature, odour, and biochemical oxygen demand and chemical oxygen demand. The magnitude of waste management issues depends on how much waste volume there is, the nature of the pollutants it carries, the rate at which it is discharged and the capacity of the receiving environment to assimilate the pollutants. Many countries dispose of such liquid wastes through their municipal sewage systems or directly into a waterway. The receiving waterbody should be able to degrade the organic and inorganic waste components in a way that does not damage the aquatic ecosystem.

Treatments can be primary and secondary.

- Primary treatments: use physical methods such as flotation, screening, and sedimentation to remove oil and grease and other suspended solids.

- Secondary treatments: use biological and physicochemical means. Biological treatments use microorganisms to metabolise the organic polluting matter into energy and biomass. "These microorganisms can be aerobic or anaerobic. The most used aerobic processes are activated sludge system, aerated lagoons, trickling filters or bacterial beds and the rotating biological contractors. In anaerobic processes, the anaerobic microorganisms digest the organic matter in tanks to produce gases (mainly methane and CO2) and biomass. Anaerobic digesters are sometimes heated, using part of the methane produced, to maintain a temperature of 30 to 35°C. In the physicochemical treatments, also called coagulation-flocculation, a chemical substance is added to the effluent to reduce the surface charges responsible for particle repulsions in a colloidal suspension, thus reducing the forces that keep its particles apart. This reduction in charge causes flocculation (agglomeration) and particles of larger sizes are settled and clarified effluent is obtained. The sludge produced by primary and secondary treatments is further processed in digesting tanks through anaerobic processes or sprayed over land as a fertilizer. In the latter case, care must be exercised to ensure that the sludge is freed of its pathogens."

Transport

Fish is transported widely in ships, and by land and air, and much fish is traded internationally. It is traded live, fresh, frozen, cured and canned. Live, fresh and frozen fish need special care.

- *Live fish:* When live fish are transported they need oxygen, and the carbon dioxide and ammonia that result from respiration must not be allowed to build up. Most fish transported live are placed in water supersaturated with oxygen (though catfish can breathe air directly through their gills and body skin, and the climbing perch has special air-breathing organs). The fish are often "conditioned" (starved) before they are transported to reduce their metabolism and increase packing density, and the water can be cooled to further reduce metabolism. Live crustaceans can be packed in wet sawdust to keep the air humid.
- *By air:* Over five percent of the global fish production is transported by air. Air transport needs special care in preparation and handling and careful scheduling. Airline transport hubs often require cargo transfers under their own tight schedules. This can influence when the product is delivered, and consequently the condition it is in when it is delivered. The air shipment of leaking seafood packages causes corrosion damage to aircraft, and each year, in the US, requires millions of dollars to repair the damage. Most airlines prefer fish that is packed in dry ice or gel, and not packed in ice.

- *By land or sea:* "The most challenging aspect of fish transportation by sea or by road is the maintenance of the cold chain, for fresh, chilled and frozen products and the optimisation of the packing and stowage density. Maintaining the cold chain requires the use of insulated containers or transport vehicles and adequate quantities of coolants or mechanical refrigeration. Continuous temperature monitors are used to provide evidence that the cold chain has not been broken during transportation. Excellent development in food packaging and handling allow rapid and efficient loading, transport and unloading of fish and fishery products by road or by sea. Also, transport of fish by sea allows for the use of special containers that carry fish under vacuum, modified or controlled atmosphere, combined with refrigeration."

Quality and Safety

The International Organisation for Standardisation, ISO, is the worldwide federation of national standards bodies. ISO defines *quality* as "the totality of features and characteristics of a product or service that bear on its ability to satisfy stated or implied needs." (ISO 8402). The quality of fish and fish products depends on safe and hygienic practices. Outbreaks of fish-borne illnesses are reduced if appropriate practices are followed when handling, manufacturing, refrigerating and transporting fish and fish products. Ensuring standards of quality and safety are high also minimises the post-harvest losses."

"The Hazard Analysis and Critical Control Points (HACCP) system of assuring food safety and quality has now gained worldwide recognition as the most cost-effective and reliable system available. It is based on the identification of risks, minimising those risks through the design and layout of the physical environment in which high standards of hygiene can be assured, sets measurable standards and establishes monitoring systems. HACCP also establishes procedures for verifying that the system is working effectively. HACCP is a sufficiently flexible system to be successfully applied at all critical stages — from harvesting of fish to reaching the consumer. For such a system to work successfully, all stakeholders must cooperate which entails increasing the national capacity for introducing and maintaining HACCP measures. The system's control authority needs to design and implement the system, ensuring that monitoring and corrective measures are put in place."

"The fishing industry must ensure that their fish handling, processing and transportation facilities meet requisite standards. Adequate training of both industry and control authority staff must be provided by support institutions, and channels for feedback from consumers established. Ensuring high standards for quality and safety is good economics, minimising losses that result from spoilage, damage to trade and from illness among consumers."

Final Products

Finfish, or parts of finfish, are typically presented physically for marketing in one of the following forms

- whole fish: the fish as it originally came from the water, with no physical processing
- drawn fish: a whole fish which has been eviscerated, that is, had its internal organs removed
- dressed fish: fish that has been scaled and eviscerated, and is ready to cook.
- pan dressed fish: a dressed fish which has had its head, tail, and fins removed, so it will fit in a pan.
- filleted fish: the "fleshy sides of the fish, cut lengthwise from the fish along the backbone. They are usually boneless, although in some fish small bones called "pins" may be present; skin may be present on one side, too. Butterfly fillets may be available. This refers to two fillets held together by the uncut flesh and skin of the belly"
- fish steaks: large dressed fish can be cut into cross section slices, usually half to one inch thick, and usually with a cross section of the backbone
- fish sticks: "are pieces of fish cut from blocks of frozen fillets into portions at least 3/8-inch thick. Sticks are available in fried form ready to heat or frozen raw, coated with batter and breaded, ready to be cooked"
- fish cakes: are "prepared from flaked fish, potatoes, and seasonings, and shaped into cakes, coated with batter, breaded, and then packaged and frozen, ready-to-be-cooked"
- fish fingers
- fish roe

Value Addition

In general value addition means "any additional activity that in one way or the other change the nature of a product thus adding to its value at the time of sale." Value addition is an expanding sector in the food processing industry, especially in export markets. Value is added to fish and fishery products depending on the requirement of different markets. Globally a transition period is taking place where cooked products are replacing traditional raw products in consumer preference.

"In addition to preservation, fish can be industrially processed into a wide array of products to increase their economic value and allow the fishing industry and exporting countries to reap the full benefits of their aquatic resources. In addition, value processes generate further employment and hard currency earnings.

This is more important nowadays because of societal changes that have led to the development of outdoor catering, convenience products and food services requiring fish products ready to eat or requiring little preparation before serving."

> *"However, despite the availability of technology, careful consideration should be given to the economic feasibility aspects, including distribution, marketing, quality assurance and trade barriers, before embarking on a value addition fish process."*

- Surimi: Surimi and surimi-based products are an example of value added products. Surimi is prepared from the mechanically deboned, washed (bleached) and stabilised flesh of fish. "It is an intermediate product used in the preparation of a variety of ready to eat seafood such as kamaboko, fish sausage, crab legs and imitation shrimp products. Surimi-based products are gaining more prominence worldwide, because of the emergence of Japanese restaurants and culinary traditions in North America, Europe and elsewhere. Ideally, surimi should be made from low-value, white fish with excellent gelling ability and which are abundant and available year-round. At present, Alaskan pollack accounts for a large proportion of the surimi supply. Other species, such as sardine, mackerel, barracuda, striped mullet have been successfully used for surimi production."
- Fishmeal and fish oil: "A significant proportion of the world catch (20 percent) is processed into fishmeal and fish oil. Fishmeal is a ground solid product that is obtained by removing most of the water and some or all of the oil from fish or fish waste. This industry was launched in the 19th century, based mainly on surplus catches of herring from seasonal coastal fisheries to produce oil for industrial uses in leather tanning and in the production of soap, glycerol and other non-food products. Presently, it uses small oily fish to produce fishmeal and oil. It is worthy to mention that, only where it is uneconomic or impracticable for human consumption, should the catch be reduced to fishmeal and oil. Indeed, cycling fish through poultry or pigs is a loss because there is a need for 3 kg of edible fish to produce approximately 1 kg of edible chicken or pork."

History

There is evidence humans have been processing fish since the early Holocene. For example, fishbones (c. 8140–7550 BP, uncalibrated) at Atlit-Yam, a submerged Neolithic site off Israel, have been analysed. What emerged was a picture of "a pile of fish gutted and processed in a size-dependent manner, and then stored for future consumption or trade. This scenario suggests that technology for fish storage was already available, and that the Atlit-Yam inhabitants could enjoy the economic stability resulting from food storage and trade with mainland sites."

Figure: A medieval view of fish processing, by Peter Brueghel the Elder (1556).

Live Fish Trade

Figure: Production line of live ornamental fish

The live fish trade can refer to the live food fish trade (for dinner) or to the ornamental fish trade (for aquariums). The fish can come from many places, but most comes from Southeast Asia.

The live food fish trade is a global system that links fishing communities with markets, primarily in Hong Kong and mainland China. Many of the fish are captured on coral reefs in Southeast Asia or the Pacific Island nations.

Consumer Demand

Within the live food trade there are certain types of fish demanded more often by consumers, particularly smaller and medium-sized fish. According to the book *While Stocks Last: The Live Reef Food Fish Trade* consumer demand has caused the fish captured on coral reefs to be the most valued fish in the trade. Consumers are important because they are directly purchasing these fish species at restaurants and stores. In addition to these types of fishes, many juvenile fish are used for the live food trade. There are also cultural and regional preferences among consumers, for example, Chinese consumers often prefer their fish to be reddish in colour believing the colour to be auspicious (Sadovy, Y.J, 393). These preferences inevitably affect the biodiversity of marine life making certain fish species rarer to find.

The life fish food trade is a lucrative business. According to University of Washington Professor Patrick Christie, live fish caught for food export earns approximately $6000 a ton. To help support themselves and their families, fishermen in Oceania and Southeast Asia sometimes use illegal fishing methods. Although many feel the fish are worth the cost, a typical dinner can cost up to one hundred dollars per kilogram. The wholesale value on these fish is anywhere from eleven US dollars to sixty-three US dollars per kilogram, meaning there's a large markup and resale value. (Hong Kong alone is estimated to be about four hundred million US dollars a year.) Because this trade frequently uses illegal methods of collecting (using cyanide), there is no way to know for sure how much money is being made each year on live fish trade, although estimates conclude probably over one billion US dollars each year.

As is often the case, consumers are willing to pay large amounts of money on rare and fresh fish. One 500-pound, polka-dot grouper, estimated to be more than a century old, was hacked into fillets by seven kitchen workers in about half an hour, the Economist reports. It was expected to bring about $15,000. (Moll 1996)

Market and Trade Routes

The centre for the Live Food Fish Trade is located in Hong Kong — the markets consumers contribute $400 million to the estimated $1 billion of the trades global value (Seaweb). Total imports flowing into Hong Kong included 10153 metric tons, of which 30 percent was re-exported to mainland China (Montaldi). Other major markets include Singapore, mainland China, and Taiwan (Graham). The primary suppliers of wild caught fish are Indonesia (accounting for nearly 50 percent of Hong Kong's imports), Thailand, Malaysia, Australia, and Vietnam (Graham). However, Taiwan and Malaysia are leading the charge towards farmed live fish specialising in an industry that "harvested annually has probably been in the billions Metric Tons (Graham)." Farming live fish is gaining popularity as tastes for live fish are burgeoning across South Asia and countries look to become more and more self sustainable, this is appareant in nations with sizable Chinese populations such as Indonesia and Malaysia.

Hong Kong and China are the dominant markets for the live fish, in addition to other cities in the region that have large Chinese populations, including Singapore and Kuala Lumpur (BusinessWorld 1 Nov. 1999). In Southeast Asia, Singapore alone consumes 500 tons of live coral fish a year (ibid). Exports from Southeast Asia rose to over 5,000 tons in 1995 from 400 tons in 1989(ibid). However, in 1996, exports declined by 22%(ibid). Indonesia, which accounts for over 60% of the harvest, saw exports falling by over 450 tons(ibid). That same year, other Southeast Asian countries have experienced similar drops in stocks of live coral fish for food(ibid). In 1996, the Philippines' exports were halved, while Malaysian exports declined by over 30%(ibid). These decreases in catch have been due to the excessive amount of fish caught for exports and the degradation of the coral reefs from such procedures.

Corruption in the trade

The live fish trade is a complex issue that involves many different perspectives, all of which must be considered in trying to approach a solution. While, at first, one may point the finger at the fishermen themselves as the criminals, there are many other factors. One is the economic disparity of many of the communities that take part in cyanide, dynamite, or other illegal fishing practices. 40% of the Filipino population and 27% of the Indonesian population is considered to be in poverty. The World Factbook) Many whose livelihood once depended on fishing or agriculture are realising it is more lucrative to participate in illegal fishing activities.

Community members who are not a part of the trade are affected by the activities of these illegal fishers. The cyanide fishers profit by taking away from everyone else's trade and food. (Lowe, 7) '*If people were using poison and my take dropped to only a little, I would accept it*,' Puah said. '*But I feel heartsick...I catch nothing at all. I have not caught a big fish in a month so there's no point in going fishing this afternoon*."' (Lowe, 7) The local people are often helpless to protect themselves, as government and law enforcement officials have "open pockets" and are also involved in the trade by turning a blind eye to the illegal actions and receiving a take of the profits. "Culpability in cyanide use cannot be understood apart from the larger structures of corruption that permeates resource extraction throughout Indonesia. The Indonesian State bureaucracy extends from Jakarta down to the village level, and radiates out into villages through kinship connections. It is the factor most tightly correlated with illegal trade in natural resources throughout Indonesia." (Lowe, 8)

Cause and Effect

Coral reefs found in the South Pacific are regarded as the "rainforest of the sea" harbouring countless fish species large and small. However, recently the live fish trade has threatened the sanctity of these endangered areas. The Global Coral Reef Monitoring Network has issued a recent report that estimates that 25% of the world's reefs are severely damaged and another third are in grave danger.

(Moore Online) The live fish trade is part of this alarming ecological trend caused by the popular use of cyanide which is injected into the coral reefs to stun inhabiting fish so they can be easily caught by nets. It is estimated that since the 1960s, more than one million kilograms of cyanide has been squirted into Philippine reefs alone, and since then the practice has spread throughout the South Pacific. (Moore Online) The live fish trade is only growing, in 1994 the Philippines exported 200,000 kg of live fish; by 2004 the Philippines were annually exporting 800,000 kg annually. (Aguiba Online) Although Asian markets are the primary buyers of live reef fish for food, the recently created U.S. Coral Reef Task Force has concluded that the U.S. is the primary purchaser of live reef fish for aquariums as well as eclectic jewellery. (Moore Online) Even though the use of cyanide in the live fish trade is severely detrimental, one must realise that this issue is multidimensional. Small-scale native fishermen of the small South Pacific coastal communities are the backbone of the live fish trade, and are forced to resort to the illegal use of sodium cyanide due to demand and high prices offered by the industry.

Fishing Techniques

While live food fish trade can be very profitable for those involved, there are many dangerous aspects to it. Through the use of illegal practices such as cyanide fishing, coral reefs and fish communities are put in grave danger. The process of cyanide fishing involves injecting crushed cyanide tablets and squirting this solution from a bottle toward the targeted fish on top of coral heads. Specifically, the cyanide kills coral polyps, symbiotic algae, and other coral reefs organisms that are necessary for maintaining the health of the coral reef. These damages eventually deteriorate the coral reef and lead it into collapse of the entire coral reef ecosystem. The effect on the targeted fish is disorientation and semi-paralysis. After being squirted with cyanide the fish is easy brought to the surface and kept alive in small, on board container. Fishermen often understand that this practice is harmful and will offer locals a portion of the fish in order to continue fishing. When ingested, small levels of cyanide accumulate in the system causing weakness of fingers and toes, failure of the thyroid gland and blurred vision. Divers without experience may come in direct contact with cyanide, causing death. Estimates show that since the 1960's, over a million kilograms of cyanide have been squirted into the coral reefs of the Philippines alone (Bryant et al.). The harm upon the reefs is coming full circle and having a social impact through the limited fish stocks. As fish are depleted from these fishing techniques, the fishermen are having a more difficult time feeding themselves (Cyanide). Additionally, the use of explosives may be used as a fishing technique in the live food fish trade. While the majority of these fish do not survive the blast of such explosions, the remaining fish that are only stunned are collected for the live food fish trade. The use of cyanide makes a stronger argument in that the more coral reef fish captured alive, the more lucrative the catch is for the fisherman. Live fish, according to the WWF, fetch five times more than a dead fish (Cyanide). This is why banning the live-fish trade would similarly harm the reefs. Fisherman would resort to the "dead fish trade", be forced to deal in a larger quantity of fish, and the process of dangerous fishing techniques continues. To illustrate the effects of repeated explosions in the proximity of a coral reef system, roughly one half of the coral reefs in Komodo National Park in Indonesia have been destroyed (Moore). The use of cyanide and explosives in fishing proves to be an effective technique in catching fish, but its powers are indiscriminating and as a result the coral reefs are being held hostage to such practices. The future of the reefs are in question just as are the futures of those who subsist from them because "from a long term perspective, the question of ethics of using the ocean...contains a commitment for future generations" (Grover 129).

Impacts on Humans

In communities like those in the Philippines and Indonesia, people are participating in the live food fish trade because it is a source of income, or at least a source of temporary income. For some communities this is one of the few income-generating opportunities. Along with the environmental, ecological, and economic consequences of this industry, there are serious health risks as well. Because of inadequate training and lack of quality equipment, divers, especially young men are in large risk of paralysis (Donaldson 2003)

Sustainable Practices

Because of the great profitability of this industry, there is a great incentive to identify sustainable practices. The Marine Aquarium Council (MAC) works to offer the hobbyist with a product that is certified as environmentally sound and sustainable. Additionally, the International Marinelife Alliance (IMA), The Nature Conservancy (TNC), and MAC are working with the Hong Kong Chamber of Seafood Merchants to develop standards for the live fish trade. The Hong Kong Seafood Merchants represent ninety percent of the buyers of live reef food fish in Hong Kong and have an extensive impact on collection practices.

Aquaculture

In an effort to address the damage inflicted on coral reef eco-systems and fish stocks, aquaculture is being utilised to reduce pressure on coral reefs. However, initial efforts to farm grouper have met with significant challenges. There are difficulties with fragile grouper seed that can make it more expensive than wild caught larvae, which can affect natural replenishment rates. Additionally, there are problems with finding suitable food, disease and cannibalism (Johannes and Ogburn). Efforts are also being made in regards to the aquarium fish trade. Juvenile fish are being captured and raised specifically for the industry. However, there are debates as to whether this practice will affect replenishment rates. "The age of the juveniles is pivotal to the debate, harvesting of postlarvae from the water column is considered to have a much lower (negligible) impact on rates of replenishment than the removal of the larger juveniles from benthic habitats because the postlarvae have yet to undergo severe mortality" (Bell, Doherty and Hair). If studies determine that the capture of juveniles is sustainable, it may help in mitigating the damage from cyanide fishing.

It should also be noted that aquaculture production, specifically grouper rearing is rapidly expanding in Asia. From 1998 to 2001 the Indo-Pacific countries involved in aquaculture; China, Indonesia, Republic of Korea, Kuwait, Malaysia, Philippines, Singapore, and Thailand witnessed a 119 percent increase in output (FAO Fishery Information Data and Statistics Unit 2003). The explosion in this practice can most likely be attributed to the large profit margins that can be derived in very little time. It is estimated that the majority of farms after annual returns can be paid back in less than one year (Siar et al. 2002). In comparison to other species of fish such as the Milkfish, the Grouper, because of high demand is able to garner high rates of return, in order to earn 1,000 dollars a grouper farm would only have to raise 400 kilograms in contrast to 5,000 kilograms of

Common Carp

The Common carp (*Cyprinus carpio*) is a widespread freshwater fish of eutrophic waters in lakes and large rivers in Europe and Asia. The wild populations are considered vulnerable to extinction, but the species has also been domesticated and introduced into environments worldwide, and is often considered an invasive species. It gives its name to the carp family: Cyprinidae. It is on the List of the world's 100 worst invasive species.

Taxonomy

- *Cyprinus carpio carpio* (European carp). Eastern Europe (notably the Danube and Volga Rivers).
- *Cyprinus carpio haematopterus* (Amur carp). Eastern Asia.
- *Cyprinus carpio rubrofuscus*. Southeastern Asia. Treated as a separate species *Cyprinus rubrofuscus* by many authorities.

It is related to the common goldfish (*Carassius auratus*), with which it is capable of interbreeding.

History

The common carp is a fish native to Asia which has been introduced to every part of the world with the exception of The Middle East and the poles. The original common carp was that found in the inland delta of the Danube river about 2000 years ago, and was torpedo-shaped and golden-yellow in colour. It had two pairs of barbels and a mesh-like scale pattern. Although this fish was initially kept as an exploited captive, it was later maintained in large, specially built ponds by the Romans in south-central Europe (verified by the discovery of common carp remains in excavated settlements in the Danube delta area). As aquaculture became a profitable branch of agriculture, efforts were made to farm the animals, and the culture systems soon included spawning and growing ponds. The common carp's native range also extends to the Black Sea, Caspian Sea and Aral Sea.

Both European and Asian subspecies have been domesticated. In Europe, domestication of carp as food fish was spread by monks between the 13th and 16th centuries. The wild forms of carp had reached the delta of the Rhine in the twelfth century already, probably also with some human help. Variants that have arisen with domestication include the mirror carp, with large mirror-like scales (linear mirror – scaleless except for a row of large scales that run along the lateral line; originating in Germany), the leather carp (virtually unscaled except near dorsal fin), and the fully scaled carp. Koi carp (&"É› (nishikigoi) in Japanese, É›Z› (pinyin: l-œ yú) in Chinese) is a domesticated ornamental variety that originated in the Niigata region of Japan in the 1820s. They also invaded the Great Lakes in 1896 when the area near Newmarket flooded and allowed them to escape into the Holland River.

Physiology

Figure: Dutch wild carp

Wild common carp are typically slimmer than domesticated forms, with body length about four times body height, red meat and a forward protruding mouth. Their growth rate by weight is about half the growth rate of domesticated carp They do not reach the length and weights of domesticated carp, which (range, 3.2–4.8 times) can grow to a maximum length of 120 centimetres (47 in), a maximum weight of over 40 kilograms (88 lb), and an oldest recorded age of at least 65 years, but reliable information seems to exist about nishikigoi of over 100 years. The largest recorded carp, caught by an angler in January 2010 at Lac de curtons (Rainbow Lake) near Bordeaux, France, weighed 42.6 kilograms (94 lb). The largest recorded weight of a domesticated carp was of a Nishikigoi especially bred for size with a length of 125 cm and a weight of 45 kg. The wild, non-domesticated forms tend to be much less stocky at around 20%–33% the maximum size.

Habitat

Figure: Carps from Vltava river, Czech Republic

Although they are very tolerant of most conditions, common carp prefer large bodies of slow or standing water and soft, vegetative sediments. A schooling fish, they prefer to be in groups of 5 or more. They naturally live in a temperate climate in fresh or slightly brackish water with a pH of 6.5–9.0 and salinity up to about 5‰, and temperatures of 3 to 35 °C. The ideal temperature is 23 to 30 °C, with spawning beginning at 17–18 °C; they will readily survive winter in a frozen over pond, as long as some free water remains below the ice. Carp are able to tolerate water with very low oxygen levels, by gulping air at the surface.

Diet

Common carp are omnivorous. They can eat a vegetarian diet of water plants, but prefer to scavenge the bottom for insects, crustaceans (including zooplankton), crawfish, and benthic worms.

Reproduction

An egg-layer, a typical adult fish can lay 300,000 eggs in a single spawning. Although carp typically spawn in the spring, in response to rising water temperatures and rain fall, carp can spawn multiple times in a season. In commercial operations spawning is often stimulated using a process called hypophysation where lyophilized pituitary extract is injected into the fish. The pituitary extract contains gonadotropic hormones which stimulate gonad maturation and sex steroid production, ultimately promoting reproduction.

Predation

A single carp can lay over a million eggs in a year, yet their population remains the same, so the eggs and young perish in similarly vast numbers. Eggs and fry often fall victim to bacteria, fungi, and the vast array of tiny predators in the pond environment. Carp which survive to juvenile are preyed upon by other fish such as the northern pike and largemouth bass, and a number of birds (including cormorants, herons, goosander, and osprey) and mammals (including otter and mink).

Introduction into Other Habitats

Common carp has been introduced, sometimes illegally, to most continents and some 59 countries. Due to their fecundity and their feeding habit of grubbing through bottom sediments for food they are notorious for altering their environment. In feeding, they may destroy, uproot, disturb and eat submerged vegetation causing serious damage to native duck and fish populations, like canvas backs. Similar to the Grass Carp, the vegetation they consume is not completely digested and rots, raising the nutritional level of the water and causing excessive algae growth. They destroy nests of other fish and eat their eggs, reducing their numbers significantly.

Efforts to non-chemically eradicate a small colony from Tasmania's Lake Crescent have been successful, however the long-term, expensive and intensive undertaking is an example of both the possibility and difficulty of safely removing the species once it is established. It has been proposed, but is regarded as environmentally questionable, to control common carp by deliberate exposure to the common carp specific Koi herpes virus with its high mortality rate. In Utah Lake the common carp's population is expected to be reduced by 75% by using nets to catch millions of them and either give them to people who will eat them or processing them into fertilizer. This in turn will give the native June sucker a chance to recover its declining population. Another method is by trapping them in tributaries they use to spawn with seine nets and exposing them to Rotenone. This method has shown to reduce their impact within 24 hours and greatly increase the native vegetation and desirable fish species. This also leaves the baby carp easily preyed upon by native fish.

In Australia there is enormous anecdotal and mounting scientific evidence that introduced carp are the cause of permanent turbidity and loss of submergent vegetation in the Murray-Darling river system, with severe consequences for river ecosystems, water quality and native fish species. In Victoria, Australia, Common carp has been declared as noxious fish species therefore there is no restriction on the quantity that a fisher can take. In South Australia, it is an offence for this species to be released back to the wild. An Australian company churns common carp into plant fertilizer.

Common carp were brought to the United States in 1831. In the late 1800s they were distributed widely throughout the country by the government as a foodfish. However, common carp are no longer prized as a foodfish in the United States. As in Australia, their introduction has been shown to have negative environmental consequences and they are usually considered to be invasive species. Millions of dollars are spent annually by natural resource agencies to control common carp populations in the United States. Common carp are believed to have been introduced into the Canadian province of British Columbia from Washington

State. They were first noted in the Okanagan Valley in 1912 as was their rapid growth in population. Carp are currently distributed in the lower Columbia (Arrow Lakes), lower Kootenay, Kettle (Christina Lake), and throughout the Okanagan system.

As Food and Sport

Cyprinus carpio is the number one fish of aquaculture. The annual tonnage of common carp, not to mention the other cyprinids, produced in China alone exceeds the weight of all other fish, such as trout and salmon, produced by aquaculture world wide. Roughly three million tonnes are produced annually, accounting for 14% of all farmed freshwater fish in 2002. China is by far the largest commercial producer, accounting for about 70% of carp production. Carp is eaten in many parts of the world both when caught from the wild and raised in aquaculture.

In Central Europe, carp is a traditional part of a Christmas Eve dinner. Many people in Poland and Czech Republic three or two days before Christmas Eve buy live carp and bring it home. Carp is kept one or two days in bathtub and then killed. Traditional Czech Christmas Eve dinner is thick soup of carp's head and offal, fried carp meat with potato salad or boiled carp in black sauce. In some Czech families the carp is not killed, but after Christmas returned to river or pond. In Western Europe, the carp is cultured more commonly as a sport fish although there is a small market as food fish. Carp are mixed with other common fish to make gefilte fish, popular in Jewish cuisine.

Common carp are extremely popular with anglers in many parts of Europe, and their popularity as quarry is slowly increasing among anglers in the United States (though destroyed as pests in many areas), and southern Canada. Carp are also popular with spear, bow, and fly fishermen.

The Romans farmed carp and this pond culture continued through the monasteries of Europe and to this day. In China, Korea and soon after in Japan carp farming took place as early as the Yayoi Period (ca. 300 B.C – 300 A.D.).

There is an increasing popularity in the United States for carp eggs that are used for caviar.

Benson (Fish)

Benson (1984 – July 2009) was "Britain's biggest and best-loved" common carp. Benson's popularity was such that she was caught 63 times in 13 years, although the accessibility that made her popular was also the cause of controversy among angling's elite. She has also been referred to as "the people's fish" and was voted by readers of *Angler's Mail* as Britain's Favourite Carp in 2005. The fish, who was female, was originally one of a pair: her original companion, Hedges, disappeared in a flood of the River Nene in 1998. Both fish were named due to a hole in Benson's dorsal fin that resembled a cigarette burn, in a reference to Benson and Hedges. At her peak weight, in 2006, she weighed 64 pounds and 2 ounces.

Benson died in July 2009, aged 25. At the time of her death, she weighed the same as a large dog and was worth £20,000. The owner of the lake where she lived alleged that she was accidentally poisoned by anglers using uncooked tigernuts as bait. Another possible cause of death was the complications during egg production.

Fame

Living in the Kingfisher Lake at the Bluebell Lakes complex, at Tansor just outside Oundle in Cambridgeshire, where she was one of approximately 150 carp; the lakes are managed "to provide the best environment for growth potential of the fish". Steve Broad, or of *UK Carp* magazine, ascribed Benson's fame to "her accessibility":

Among keen anglers there are about only 20 carp that can be seriously called "household names". Benson was near the top of that league. The thing that made Benson famous was her accessibility. Unlike other big carp, she was a day-ticket fish — anyone could go along and try to catch her.

However, this very accessibility made the fish controversial among the sport's elite: "Everyday anglers loved her because there was a chance they could have their photo taken with one of the big fish ... some serious anglers did not like her because she was open to everyone."

Benson's record of being caught so often masks her unpredictability. "There was a period when Benson was caught every Monday for six weeks. Then it seemed that she disappeared for the next 12 months."

Death

A quantity of uncooked nuts, which are toxic to fish who swell up because they cannot process them, were found nearby on the bank. Owner of Bluebell Lakes Tony Bridgefoot, 53, said he feared the fish had been killed by "irresponsible anglers" ... "It seems her demise was caused by the introduction of foods that are harmful to fish."

Benson's successor as a popular and very large common carp may not live too far away from the fish's former haunt. "The same complex where Benson lived boasts a lot of promising 40lb fish. There's one — the Z-Fish — that is ounces under 50lb and still growing."

It has since been confirmed that the most likely cause of death was not nut poisoning, but rather reproductive complications due to gravidity.

Community Supported Fishery

A community supported fishery (CSF) is a shore-side community of people collaborating with the local fishing community. Tailored after the community supported agriculture model, a CSF contributes freshly caught local seafood to the local markets while providing fishermen with a better price on less catch. CSF members give the fishing community financial support in advance of the season, and in turn the fishermen provide a weekly share of seafood during the harvesting season. Community supported fisheries aim to reconnect people with the ocean that sustains them and build a rewarding relationship between the fishermen and the shareholders.

Port Clyde Pilot

A pilot CSF was started in the fall of 2007 when the Mid-Coast Fishermen's Cooperative out of Port Clyde, Maine, the Northwest Atlantic Marine Alliance and the First Universalist Church in Rockland, Maine, created the first CSF in New England. The Port Clyde CSF began by delivering shrimp. The pilot proved so successful that in the spring of 2008, the Port Clyde community expanded their CSF beyond shrimp. Today, this program delivers fresh, Maine caught shrimp and groundfish to several different locations throughout their local community. In addition, the Midcoast Fishermen's Coop has opened a processing centre where they now pick shrimp and offer filleted fish as well.

Triple Bottom Line

Community supported fishery programs include a triple bottom line

- Environmental stewardship: to encourage an ethic of ecological stewardship that results in creative, community-based approaches to marine conservation.
- Local economies: to increase the viability of traditional coastal communities by fostering economic opportunities that support natural resource-based livelihoods.
- Social improvements: to cultivate ties and establish bonds between shoreside communities and inshore urban, suburban and rural communities by providing fresh, local seafood.

Organisations

- Northwest Atlantic Marine Alliance (NAMA)
- Midcoast Fishermen's Association
- Island Institute
- Gloucester Fishermen's Wives Association
- Penobscot East Resource Centre

- Women of Fishing Families
- Ocracoke Seafood Ocracoke, NC
- Ocracoke Working Watermen's Association Ocracoke, NC
- LocalCatch.org

Artisan Fishing

Artisan fishing is a term used to describe small-scale low-technology commercial or subsistence fishing practices. The term particularly applies to coastal or island ethnic groups using traditional techniques such as rod and tackle, arrows and harpoons, throw nets and drag nets, and traditional fishing boats. It does not usually cover the concept of fishing for sport, and might be used when talking about the pressures between large-scale modern commercial fishing practises and traditional methods, or when aid programs are targeted specifically at fishing at or near subsistence levels.

Artisan fishing is often, but not always, less intensive and less stressful on fish populations than modern industrial fishing techniques. It is subject to difficulties in the export process due to inadequate investment in refrigeration and processing facilities. However, the most important goal of artisan fishing is domestic consumption, as it is often an important source of inexpensive and accessible protein in poor coastal areas.

Overview

The world's bottom billion are falling behind in the war on poverty and hunger. Traditional staple foods alone cannot feed the world while fisheries and aquaculture have the potential to provide better nutrition and improve incomes for the poor.

Fisheries are a source of high-protein food: they provide over 1.5 billion people with 20 percent of their animal protein, and 3 billion people with 15 percent of their animal protein. In some countries they provide up to 70% of animal protein, which is particularly important for pregnant women and young children.

Catching, processing and trading fish provides a livelihood for millions of men and women, the overwhelming majority of whom are associated with small-scale fisheries. In fact, of the 70% of the world's total fish catch that comes from developing countries, over a half of this comes from small-scale fisheries.

The Challenges

Small-scale fisheries have failed to keep pace with demand for food and employment in developing countries and urgently require attention to technology, health, economics and reform in management and governance. Globally there are 38 million full-time fishers, and 20% of these fishers earn less than US$1 per day.

Most wild-fish stocks are near the limit of their productive capacity or have destabilised or declined because of overfishing and other causes. Given these pressures and constraints, the capacity to maintain wild fisheries needs to increase, and aquaculture needs further development since it is the only alternative way fish production can be substantially increased.

Climate change poses additional challenges affecting the seasonality and productivity of fisheries, as well as the geographical locations and viability of the supporting habitats while a further global threat comes from degradation of coral reefs. Seventy-five percent of coral reefs are currently under immediate and direct threat from local sources, which include overfishing, destructive fishing, coastal development, and pollution.

Fishing Village

A fishing village is a village, usually located near a fishing ground, with an economy based on catching fish and harvesting seafood. The continents and islands around the world have coastlines totalling around 356,000 kilometres (221,000 mi). From Neolithic times, these coastlines, as well as the shorelines of inland lakes and the banks of rivers, have been punctuated with fishing villages. Most surviving fishing villages are traditional.

Characteristics

Coastal fishing villages are often somewhat isolated, and sited around a small natural harbour which provides safe haven for a village fleet of fishing boats. The village needs to provide a safe way of landing fish and securing boats when they are not in use. Fishing villages may operate from a beach, particularly around lakes. For example, around parts of Lake Malawi, each fishing village has its own beach. If a fisherman from outside the village lands fish on the beach, he gives some of the fish to the village headman. Village fishing boats are usually characteristic of the stretch of coast along which they operate. Traditional fishing boats evolve over time to meet the local conditions, such as the materials available locally for boat building, the type of sea conditions the boats will encounter, and the demands of the local fisheries.

Some villages move out onto the water itself, such as the floating fishing villages of Ha Long Bay in Vietnam, the stilt houses of Tai O built over tidal flats near Hong Kong, and the kelong found in waters off Malaysia, the Philippines and Indonesia. Other fishing villages are built on floating islands, such as the Phumdi on Loktak Lakein India, and the Uros on Lake Titicaca which borders Peru and Bolivia.

Apart from catching fish, fishing villages often support enterprises typically found in other types of village, such as village crafts, transport, schools and health clinics, housing and community water supplies. In addition, there are enterprises that are natural to fishing villages, such as fish processing and marketing, and the building and maintenance of boats. Until the 19th century, some villagers supplemented their incomes with wrecking (taking valuables from nearby shipwrecks) and smuggling.

In less developed countries, some traditional fishing villages persist in ways that have changed little from earlier times. In more developed countries, traditional fishing villages are changing due to socioeconomic factors like industrial fishing and urbanisation. Over time, some fishing villages outgrow their original function as artisanal fishing villages. Seven hundred years ago, Shanghai, beside the Yangtze River delta, was a small fishing village. In recent times, fishing villages have been increasingly targeted for tourist and leisure enterprises. Recreational fishing and leisure boat pursuits can be big business these days, and traditional fishing villages are often well positioned to take advantage of this. For example, Destin on the coast of Florida, has evolved from an artisanal fishing village into a seaside resort dedicated to tourism with a large fishing fleet of recreational charter boats. The tourist appeal of fishing villages has become so big that the Korean government is purpose-building 48 fishing villages for their tourist drawing power. In 2004 China reported it had 8,048 fishing villages.

Early Villages

Skara Brae on the west coast of the Orkney mainland, off Scotland, was a small Neolithic agricultural and fishing village with ten stone houses. It was occupied from about 3100 to 2500 BC, and is Europe's most complete Neolithic village. The ancient Lycian sunken village of Kaleköy in Turkey, dates from 400 BCE. Clovelly, a fishing hamlet north Devon coast of England, an early Saxon settlement, is listed in the Domesday Book. Kaunolu Village, a Hawaiian fishing village, is thought to date from about 1500 CE.

Recent archaeological excavations of earlier fishing settlements are occurring at some pace. A fishing village recently excavated in Khanh Hoa, Vietnam, is thought be about 3,500 years old. Excavations on the biblical fishing village Bethsaida, on the shore of the Sea of Galilee and birthplace of the apostles Peter, Philip and Andrew, have shown that Bethsaida was established in the tenth century BCE. A Tongan fishing village, recently excavated, appears to have been founded 2900 years ago. This makes it the oldest known settlement in Polynesia. Another recent excavation has been made of Walraversijde, a medieval fishing village on the coast of West Flanders in Holland.

Marine Life Protection Act

The **Marine Life Protection Act** (MLPA) was passed in 1999 and is part of the California Fish and Game Code. The MLPA requires California to reevaluate all existing marine protected areas (MPAs) and potentially design new MPAs that together function as a statewide network. The MLPA has clear guidance associated with the development of this MPA network. MPAs are developed on a regional basis with MLPA and MPA specific goals in mind, and are evaluated over time to assess their effectiveness for meeting these goals.

Unlike terrestrial conservation, marine conservation often lacks a systematic approach to conserving biodiversity. Little gap analysis has been performed on the marine environment, and there is a lack of knowledge into what is protected, what needs to be protected, and where the protection needs to occur. Over the last century there has been a rapid increase in the loss of marine biodiversity and habitat degradation. About 70% of California's population lives within one hour of the coast and the ocean provides resources to local, state, and national interests. As a result, species and habitat loss has become a major issue. Over 90% of California's coastal wetlands have been lost, coastal waters have become contaminated with a variety of urban and agricultural toxins, and a large number of targeted species have declined in the last 10–20 years. Over the last two decades, California fish catches have decreased by over 50%. These impacts have decreased the health and value of the California's coastal ocean and imply a need for a more systematic approach to marine conservation. Although there is no single solution to conserving the marine environment, MPAs are a potentially valuable tool for marine conservation when designed and managed effectively. A well designed and managed network of MPAs helps to prevent degradation, fosters marine biodiversity, and may maintain a more sustainable fishing industry. The MLPA helps to promote a shift from single-species management to an ecosystem based management and is a more systematic approach to marine conservation.

A Brief History of California MPAs

California's first six MPAs were created between 1909 and 1913; by 1950 all had been removed. After 1950 more than 50 other MPAs were created along the California coast. But these MPAs were established in a random manner and without regard to regional conservation goals. Most have been thought to be too small and ineffective in protecting against habitat and species loss. With these existing MPAs less than 1% of coastal waters were protected, and none extended to deeper waters. In 1999 the MLPA was created in order to re-evaluate the current MPA system and to establish a better network of MPAs that would be more effective in protecting against habitat and species loss.

The Marine Life Protection Act

MLPA Findings

The MLPA found that existing MPAs were not created under a coherent plan or scientific guidelines, and that there is a need to redesign the MPA system. Coastal development, water pollution, and other human activities are a threat to California's diverse coastal waters. These coastal waters, along with the ecosystems and species which thrive within them are vital assets to the state and nation. An improved MPA system would help protect against habitat and ecosystem loss, conserve biological diversity, provide safe breeding grounds for fish and other marine species, improve research opportunities, create a reference point from which the rest of the ocean can be compared against, and may help to re-grow depleted fisheries.

MPA Network

The MLPA appointed the California Department of Fish and Game (CDFG) with the task of developing and managing a network of MPAs. The CDFG determines the final location and size of each MPA. The goal is to establish a network of MPAs that work together. This network takes into account the movement of adult and larval fish and also focuses on deepwater habitats for the first time. A proportion of the MPA network is to be designated as no-take zones. No-take zones allow for a large area of safe breeding grounds and a sanctuary for large, female fish. Large female fish produce more viable offspring and are vital in a population. With this idea, the MPA network has the potential to boost fish populations in areas out side of MPAs. Fishery growth has been successful along the Great Barrier Reef Marine Park and the Florida Keys National Marine Sanctuary after reserves were established in these areas. The final decision of the size and location of the MPAs depends on the species and habitats effected, stakeholder and conservation goals, and how each individual MPA will function on its own and as part of the network.

MLPA Implementation

After its passage in 1999, the CDFG began to implement the MLPA. The first attempt involved a Master Plan Team which included primarily scientific experts and governmental agencies, with little input from local stakeholders. This plan failed once it was brought to the public for approval, mostly because stakeholders and other members of the public were excluded from the process. Commercial and recreational fishers showed the most resistance, stating that MPAs produce no benefits for fisheries and objecting to the size and location of the proposed MPAs. In 2002, the CDFG implemented the MLPA for a second time. This plan involved members from the Master Plan Team, as well as seven Working Stakeholder Groups, which included governmental agency officials, recreational and commercial fishing interests, recreational divers, ocean vessel representatives, environmental interests, charter boat operators, harbourmasters, and scientists/educators. This attempt was more successful and gained public support, but the project lost funding in 2003 due to a poor fiscal year.

In 2004 the CDFG gained new funding from several organisations to initiate the Marine Life Protection Act Initiative. The Initiative divided the coast into sequential regions and assembled a Blue Ribbon Task Force on Marine Protected Areas, Science Advisory Team, and Regional Stakeholder Group to develop and evaluate the first set of MPAs in the Central Coast region. On April 13, 2007, after nearly three years of public meetings and proposal reviews, the Fish and Game Commission evaluated and voted on a final MPA proposal for the Central California coast. The commission voted on a plan to establish 29 MPAs covering approximately 204 square miles (18%) of state waters with 85 square miles (8%) designated as no-take state marine reserves. The network ranges from Pigeon Point in San Mateo County south to Point Conception in Santa Barbara County, and contains several types of MPAs with varying degrees of protection.

The Central Coast plan has received high marks for scientific effectiveness. Local stakeholders developed a balanced network that protects the region's best habitat, including parts of the Big Sur Coast and Monterey Bay, while allowing continued access to most recreational and commercial fishing grounds. California's Central Coast MPAs went into effect in September 2007 and scientific baseline data has been collected over the last two years..

The North Central Coast plan, adopted by the Fish and Game Commission on August 5, 2009, also represented a compromise between different interest groups, and protected iconic sites like the Farallon Islands, Point Reyes Headlands, and Bodega Head while leaving nearly 90 percent of coastal waters open for fishing. Regulations for the North Central Coast MPAs, which extend from Alder Creek, near Pt. Arena in Mendocino County, to Pigeon Point in San Mateo County, went into effect on May 1, 2010. The regulations established 21 marine protected areas (MPAs), three State Marine Recreational Management Areas, and six special closures, in total covering approximately 153 square miles (20.1%) of state waters in the north central coast study region. Approximately 86 square miles (11%) of the 153 square miles (400 km^2) are designated as "no take" state marine reserves, while different take allowances providing varying levels of protection are designated for the rest.

In 2008, South Coast Regional Stakeholders began a public planning process to design the part of the statewide MPA network that spans from Pt. Conception in Santa Barbara county to the U.S. border with Mexico. On Dec. 15, 2010 the CA Fish and Game Commission adopted regulations to create 36 new MPAs encompassing approximately 187 square miles (8 percent) of state waters in the study region. Approximately 116 square miles (4.9 percent) have been designated as no-take state marine reserves (82.5 square miles/3.5 percent) and no-take state marine conservation areas (33.5 square miles/1.4 percent), with the remainder designated as state marine conservation areas with different take allowances and varying levels of protection. Implementation of the new South Coast MPAs will take place beginning October 1, 2011.

The North Coast region, which stretches from Point Arena to the Oregon border, concluded the stakeholder planning process in August 2010. Stakeholders developed a unified proposal for their regional MPA network supported by fishermen, conservationists, and tribal representatives. The unified plan will be considered by the Fish and Game Commission beginning in 2011.

The Science

An extensive body of peer-reviewed studies on marine protected areas have concluded that well-designed networks of protected waters are effective in improving ocean health and making ocean waters more resilient. Most recently, the February 2010 issue of the Proceedings of the National Academy of Sciences (PNAS) included several new studies that showed that scientifically-based MPA networks have a net positive impact on both ecosystem productivity and associated fisheries. One of the studies found that such well-designed networks can simultaneously improve the quality of ocean habitat, increase the size and abundance of sea life, and increase fishing yields and profits. Several studies have stressed the importance of location: The location of protected areas is important. In order to be effective, marine reserves must be placed in the areas where fish and shellfish feed and breed.

MLPA Controversy / Conflict of Interest Investigation

The MLPA Initiative recently began to receive negative press from fishing rights groups and individual fishermen due to the apparent conflict of interest of some MPA officials and unfair practices in the MLPA process. On May 14, 2009, Fish and Game Commission Member Jim Kellogg called for the Commission to put the MLPA process on hold due to the state budget crisis. Senate Majority Leader Dean Florez and North Coast Assemblyman Wes Chesbro requested investigation into conflicts of interest among Blue Ribbon Task Force members as well as sources of funding for the MLPA.

On May 19, 2009, the California Fair Political Practices Commission (CFPPC) disclosed that the Enforcement Division of the Fair Political Practices Commission has started a formal investigation into Fish and Game Commissioner George Michael Sutton, under charge of having violated the Political Reform Act (PRA) of 1974 due his conflicts of interest on votes on the MLPA while serving on the Fish and Game Commission. On June 24, after investigating the matter, the Fair Political Practices Commission (FPPC) declared that Commissioner Sutton can participate in any and all public processes and deliberations surrounding the Marine Life Protection Act (MLPA) without conflict.

Lawsuits

On January 27, 2011, a number of member organisations of the Partnership for Sustainable Oceans led by Robert C. Fletcher (a former California Department of Fish and Game Chief Deputy director) filed a lawsuit challenging closures of certain fishing areas under the California Fish and Game Commission (CGFC). The plaintiff's challenged the legality of the CGFC's ability to impose its own regulation and claimed that the commission had violated the state's Administrative Procedures Act and the California Environmental Quality Act.

As part of administrative state government process, the Game and Fish Commission is required by law to disclose to the public all matters concerning MLPA research, regulation and closures. The BRTF (Blue Ribbon Task Force) and MPT (Master Plan Team), who were charged with the task of conducting research and establishing a conservation plan, had been executively ordered not to disclose the information it had obtained and produced by order of California Resource Secretary Mike Chrisman. Documents obtained in the PRA lawsuit included an email dated April 7, 2007 and advised BRTF members to "give your own notes verbally and throw them away after." This constituted a breach of the California Public Records Act, which states that "the people have the right of access to information concerning the conduct of the people's business".

On March 10, 2011, a California Superior Court ruling found in favour of the plaintiffs citing that the BRTF and MPT had failed to produce evidence that it complied with the California Public Records act while conducting research from April 2007 through November 2009. It cited that the practices used by the privately funded BRTF and MPT under the GFC did not comply with "open and transparent" processes as outlined in the Public Records Act and ordered the California Game and Fish Commission to pay all legal fees incurred by Mr. Fletcher and his team.

7

Aquarium Management

An aquarium (plural *aquariums* or *aquaria*) is a vivarium consisting of at least one transparent side in which water-dwelling plants or animals are kept. Fishkeepers use aquaria to keep fish, invertebrates, amphibians, marine mammals, turtles, and aquatic plants. The term combines the Latin root *aqua*, meaning water, with the suffix *-arium*, meaning "a place for relating to".

An aquarist owns fish or maintains an aquarium, typically constructed of glass or high strength acrylic plastic. Cuboid aquaria are also known as fish tanks or simply tanks, while bowl-shaped aquaria are also known as fish bowls. Size can range from a small glass bowl to immense public aquaria. Specialised equipment maintains appropriate water quality and other characteristics suitable for the aquarium's residents.

History and Popularisation

Antiquity

In the Roman Empire, the first fish to be brought indoors was the sea barbel, which was kept under guest beds in small tanks made of marble. Introduction of glass panes around the year 50 allowed Romans to replace one wall of marble tanks, improving their view of the fish.

In 1369, the Chinese Emperor, HóngwÔ, established a porcelain company that produced large porcelain tubs for maintaining goldfish; over time, people produced tubs that approached the shape of modern fish bowls. Leonhard Baldner, who wrote *Vogel-, Fisch- und Tierbuch* (Bird, Fish, and Animal Book) in 1666, maintained weather loaches and newts.

Nineteenth Century

In 1836, soon after his invention of the Wardian case, Dr. Nathaniel Bagshaw Ward proposed to use his tanks for tropical animals. In 1841 he did so, though only with aquatic plants and toy fish. However, he soon housed real animals. In 1838, Félix Dujardin noted owning a saltwater aquarium, though he did not use the term. In 1846, Anne Thynne maintained stony corals and seaweed for almost three years, and was cred as the creator of the first balanced marine aquarium in London. At about the same time, Robert Warington experimented with a 13-gallon container, which contained goldfish, eelgrass, and snails, creating one of the first stable aquaria. He published his findings in 1850 in the Chemical Society's journal.

The keeping of fish in an aquarium became a popular hobby and spread quickly. In the United Kingdom, it became popular after ornate aquaria in cast iron frames were featured at the Great Exhibition of 1851. In 1853, the first large public aquarium opened in the London Zoo and came to be known as the Fish House. Philip Henry Gosse was the first person to actually use the word "aquarium", opting for this term (instead of "aquatic vivarium" or "aqua-vivarium") in 1854 in his book *The Aquarium: An Unveiling of the Wonders of the Deep Sea*. In this book, Gosse primarily discussed saltwater aquaria. In the 1850s, the aquarium became a fad in the United Kingdom. Tank designs and techniques for maintaining water quality were developed by Warington, later cooperating with Gosse until his critical review of the tank water composition. Edward Edwards developed these glass-fronted aquaria in his 1858 patent for a "dark-water-chamber slope-back tank", with water slowly circulating to a reservoir beneath.

Germans soon rivalled the British in their interest. In 1854, an anonymous author had two articles published about the saltwater aquaria of the United Kingdom: *Die Gartenlaube* (The Garden House) entitled *Der Ocean auf dem Tische*. However, in 1856,*Der See im Glase* (The Lake in a Glass) was published, discussing freshwater aquaria, which were much easier to maintain in landlocked areas. During the 1870s, some of the first aquarist societies were appearing in Germany. The United States soon followed. Published in 1858, Henry D. Butler's *The Family Aquarium* was one of the first books written in the United States solely about the aquarium. According to the July issue of *The North American Review* of the same year, William Stimson may have owned some of the first functional aquaria, and had as many as seven or eight. The first aquarist society in the United States was founded in New York City in 1893, followed by others. The *New York Aquarium Journal*, first published in October 1876, is considered to be the world's first aquarium magazine.

In the Victorian era in the United Kingdom, a common design for the home aquarium was a glass front with the other sides made of wood (made watertight with a pitch coating). The bottom would be made of slate and heated from below. More advanced systems soon began to be introduced, along with tanks of glass in metal frames. During the latter half of the 19th century, a variety of aquarium designs were explored, such as hanging the aquarium on a wall, mounting it as part of a window, or even combining it with a birdcage.

Twentieth Century

Circa 1908, the first mechanical aquarium air pump was invented, powered by running water, instead of electricity. The introduction of the air pump into the hobby is considered by several historians of the hobby to be a pivotal moment in its development.

Aquaria became more widely popular as houses had an electricity supply after World War I. Electricity allowed artificial lighting as well as aeration, filtration, and heating of the water. Initially, amateur aquarists kept native fish (with the exception of goldfish); the availability of exotic species from overseas further increased the popularity of the aquarium. Jugs made from a variety of materials were used to import fish from overseas, with a bicycle foot pump for aeration. Plastic shipping bags were introduced in the 1950s, making it easier to ship fish. The eventual availability of air freight allowed fish to be successfully imported from distant regions. In the 1960s metal frames made marine aquaria almost impossible due to corrosion, but the development of tar and silicone sealant allowed the first all-glass aquaria made by Martin Horowitz in Los Angeles, CA. The frames remained, however, though purely for aesthetic reasons.

In the United States, as of 1996, aquarium keeping is the second-most popular hobby after stamp collecting. In 1999 it was estimated that over nine million U.S. households own an aquarium. Figures from the 2005/2006 APPMA National Pet Owners Survey report that Americans own approximately 139 million freshwater fish and 9.6 million saltwater fish. Estimates of the numbers of fish kept in aquaria in Germany suggest at least 36 million. The hobby has the strongest following in Europe, Asia, and North America. In the United States, 40 percent of aquarists maintain two or more tanks.

Design

Materials

Most aquaria consist of glass panes bonded together by silicone, with plastic frames that are attached to the upper and lower edges for decoration. The glass aquarium is standard for sizes up to about 1,000 litres (260 US gal). However, glass as a material is brittle and has very little give before fracturing, though generally the sealant fails first. Aquaria come in a variety of shapes such as cuboid, hexagonal, angled to fit in a corner (L-shaped), and bow-front (the front side curves outwards). Fish bowls are generally either made of plastic or glass, and are either spherical or some other round configuration in shape. The very first modern aquarium made of glass was developed in the 19th century by Robert Warrington. During the Victorian age it was common for glass aquaria to have slate or steel bottoms, which allowed the aquaria to be heated underneath with an open flame heat source. The aquaria in those days had the glass panels attached with metal frames and sealed with putty. These metal framed aquaria were still available on the market until the mid 1960s when the modern, silicone-sealed style replaced them. Acrylic tanks were not generally available to the public until the 1970s.

Although glass aquaria are usually preferred by aquarists over the acrylic ones because of their resistance to scratching and much more accessible price, they come with several disadvantages. Not only are they not as crack resistant as acrylic tanks but they are also nearly two times heavier than the latter. They also provide less insulation than acrylic aquaria and do not come in as many interesting shapes. Many aquarists or beginners who want to get fish as pets find it particularly onerous that many online suppliers will not ship glass aquaria because of the high potential for cracking and the high weight, which increases the cost of shipping. However, glass tanks are more convenient for other aquarists because unlike acrylic, glass does not yellow over time, and also because glass tanks do not need as much support as acrylic aquaria.

Even though the price is one of the main considerations for aquarists when deciding which of these two types of aquaria to purchase, when it comes to very large tanks the price difference tends to disappear.

Acrylic aquaria are also available and are the primary competitor with glass. Acrylic aquaria are stronger than glass, and much lighter. Acrylic-soluble cements are used to directly fuse acrylic together (as opposed to simply sealing the seam). Acrylic allows for the formation of unusual shapes, such as the hexagonal tank. Compared to glass, acrylics are easy to scratch; but unlike glass, it is possible to polish out scratches in acrylic.

Laminated glass is sometimes used, which combines the advantages of both glass and acrylic.

Large aquaria might instead use stronger materials such as fiberglass-reinforced plastics. However, this material is not transparent. Reinforced concrete is used for aquaria where weight and space are not factors. Concrete must be coated with a waterproof layer to prevent the water from breaking down the concrete as well as preventing contamination of the water by the concrete.

Styles

Aquariums have been fashioned into coffee tables, sinks, and even toilets. Another such example is the MacQuarium, an aquarium made from the shell of an Apple Macintosh computer. In recent years, elaborate custom-designed home aquariums costing hundreds of thousands of dollars have become status symbols—according to *The New York Times*, "among people of means, a dazzling aquarium is one of the last sure fire ways to impress their peers."

A kreisel tank is a circular aquarium designed to hold delicate animals such as jellyfish. These aquariums provide slow, circular water flow with a bare minimum of interior hardware, to prevent delicate animals from becoming injured by pumps or the tank itself. Originally a German design (*kreisel* means spinning top), the tank has no sharp corners, and keeps the housed animals away from the plumbing. Water moving into the tank gives a gentle flow that keeps the inhabitants suspended, and water leaving the tank is covered by a delicate screen that prevents the inhabitants from getting stuck. There are several types of kreisel tanks. In a true kreisel, a circular tank has a circular, submerged lid. Pseudokreisels have a curved bottom surface and a flat top surface, similar to the shape of either a "U" or a semicircle. Stretch kreisels or Langmuir kreisels are a "double gyre" kreisel design, where the tank length is at least twice the height. Using two downwelling inlets on both sides of the tank lets gravity create two gyres in the tank. A single downwelling inlet may be used in the middle as well. The top of a stretch kreisel may be open or closed with a lid. There may also be screens about midway down the sides of the tank, or at the top on the sides. It is possible to combine these designs; a circular shaped tank is used without a lid or cover, and the surface of the water acts as the continuation of circular flow. It is now possible to start a jellyfish aquarium at home as easily as a regular fish tank.

Another popular setup is the biotope aquarium. A biotope aquarium is a recreation of a specific natural environment. Some of the most popular biotopes (to name only a few) are the Amazon river, Rio Negro River, Lake Malawi, Lake Tanganyika, and Lake Victoria. The fish, plants, substrate, rocks, wood, and any other component of the display should match that of the natural environment. It can be a real challenge to recreate such environments and most "true" biotopes will only have a few species of fish (if not only one) and invertebrates.

Aquarium Size and Volume

An aquarium can range from a small glass bowl containing less than 1 litre (2.1 US pt) of water to immense public aquaria that house entire ecosystems such as kelp forests. Relatively large home aquaria resist rapid fluctuations of temperature and pH, allowing for greater system stability.

Unfiltered bowl-shaped aquaria are now widely regarded as unsuitable for most fish. Advanced alternatives are now available. In order to keep water conditions at suitable levels, aquariums should contain at least two forms of filtration: biological and mechanical. Chemical filtration should also be considered under some circumstances for optimum water quality. Chemical filtration is frequently achieved via activated carbon, to filter medications, tannins, and/or other known impurities from the water.

Reef aquaria under 100 litres (26 US gal) have a special place in the aquarium hobby; these aquaria, termed nano reefs (when used in reefkeeping), have a small water volume.

Practical limitations, most notably the weight of water (1 kilogram per litre (8.345 lb/U.S. gal)) and internal water pressure (requiring thick glass siding) of a large aquarium, restrict most home aquaria to a maximum of around 1 cubic metre in volume (1000 L, weighing 1,000 kg or 2,200 lb). Some aquarists, however, have constructed aquaria of many thousands of litres.

Public aquariums designed for exhibition of large species or environments can be dramatically larger than any home aquarium. The Georgia Aquarium, for example, features an individual aquarium of 6,300,000 US gallons (24,000 m^2).

Components

The typical hobbyist aquarium includes a filtration system, an artificial lighting system, and a heater or chiller depending on the aquarium's inhabitants. Many aquaria incorporate a hood, to decrease evapouration and prevent fish from leaving the aquarium (and anything else from entering the aquarium). They also often hold lights.

Combined biological and mechanical aquarium filtration systems are common. These either convert ammonia to nitrate (removing nitrogen at the expense of aquatic plants), or to sometimes remove phosphate. Filter media can house microbes that mediate nitrification. Filtration systems are the most complex component of home aquaria.

Aquarium heaters combine a heating element with a thermostat, allowing the aquarist to regulate water temperature at a level above that of the surrounding air, whereas coolers and chillers (refrigeration devices) are for use anywhere, such as cold water aquaria, where the ambient room temperature is above the desired tank temperature. Thermometres used include glass alcohol thermometres, adhesive external plastic strip thermometres, and battery-powered LCD thermometres. In addition, some aquarists use air pumps attached to airstones or water pumps to increase water circulation and supply adequate gas exchange at the water surface. Wave-making devices have also been constructed to provide wave action.

An aquarium's physical characteristics form another aspect of aquarium design. Size, lighting conditions, density of floating and rooted plants, placement of bog-wood, creation of caves or overhangs, type of substrate, and other factors (including an aquarium's positioning within a room) can all affect the behaviour and survival of tank inhabitants. An aquarium can be placed on an aquarium stand. Because of the weight of the aquarium, a stand must be strong as well as level. A tank that is not level may distort, leak, or crack. These are often built with cabinets to allow storage, available in many styles to match room decor. Simple metal tank stands are also available. Most aquaria should be placed on polystyrene to cushion any irregularities on the underlying surface or the bottom of the tank itself. However, some tanks have an underframe making this unnecessary.

Aquarium Maintenance

Large volumes of water enable more stability in a tank by diluting effects from death or contamination events that push an aquarium away from equilibrium. The bigger the tank, the easier such a systemic shock is to absorb, because the effects of that event are diluted.

For example, the death of the only fish in a three U.S. gallon tank (11 L) causes dramatic changes in the system, while the death of that same fish in a 100 U.S. gallon (400 L) tank with many other fish in it represents only a minor change. For this reason, hobbyists often favour larger tanks, as they require less attention.

Several nutrient cycles are important in the aquarium. Dissolved oxygen enters the system at the surface water-air interface. Similarly, carbon dioxide escapes the system into the air. The phosphate cycle is an important, although often overlooked, nutrient cycle. Sulphur, iron, and micronutrients also cycle through the system, entering as food and exiting as waste. Appropriate handling of the nitrogen cycle, along with supplying an adequately balanced food supply and considered biological loading, is enough to keep these other nutrient cycles in approximate equilibrium.

An aquarium must be maintained regularly to ensure that the fish are kept healthy. Daily maintenance consists of checking the fish for signs of stress and disease. Also, aquarists must make sure that the water has a good quality and it is not cloudy or foamy and the temperature of the water is appropriate for the particular species of fish that live in the aquarium.

Typical weekly maintenance includes changing around 10-20% of the water while cleaning the gravel, or other substrate if the aquarium has one. A good habit is to remove the water being replaced by "vacuuming" the gravel with suitable implements, as this will eliminate uneaten foods and other residues that settle on the substrate. Tap water is not considered to be safe for fish to live in because it contains chemicals that harm the fish, so any tap water used must be treated with a suitable water conditioner, such as a product which removes chlorine and chloramine, and neutralises any heavy metals present. The water parameters must be checked both in the tank and in the replacing water, to make sure they are suitable for the species of fish kept.

Water Conditions

The solute content of water is perhaps the most important aspect of water conditions, as total dissolved solids and other constituents dramatically impact basic water chemistry, and therefore how organisms interact with their environment. Salt content, or salinity, is the most basic measure of water conditions. An aquarium may have freshwater (salinity below 500 parts per million), simulating a lake or river environment; brackish water (a salt level of 500 to 30,000 PPM), simulating environments lying between fresh and salt, such as estuaries; and salt water or seawater (a salt level of 30,000 to 40,000 PPM), simulating an ocean environment. Rarely, higher salt concentrations are maintained in specialised tanks for raising brine organisms.

Saltwater is typically alkaline, while the pH (alkalinity or acidicity) of fresh water varies more. Hardness measures overall dissolved mineral content; hard or soft water may be preferred. Hard water is usually alkaline, while soft water is usually neutral to acidic. Dissolved organic content and dissolved gases content are also important factors.

Home aquarists typically use tap water supplied through their local water supply network to fill their tanks. Straight tap water cannot be used in localities that pipe chlorinated water. In the past, it was possible to "condition" the water by simply letting the water stand for a day or two, which allows the chlorine time to dissipate. However, chloramineis now used more often and does not leave the water as readily. Additives formulated to remove chlorine or chloramine are often all that is needed to make the water ready for aquarium use. Brackish or saltwater aquaria require the addition of a commercially available mixture of salts and other minerals.

Some aquarists modify water's alkalinity, hardness, or dissolved content of organics and gases, before adding it to their aquaria. This can be accomplished by additives, such as sodium bicarbonate, to raise pH. Some aquarists filter or purify their water through deionisation or reverse osmosis prior to using it. In contrast, public aquaria with large water needs often locate themselves near a natural water source (such as a river, lake, or ocean) to reduce the level of treatment. Some hobbyists use an algae scrubber to filter the water naturally.

Water temperature determines the two most basic aquarium classifications: tropical versus cold water. Most fish and plant species tolerate only a limited temperature range; tropical aquaria, with an average temperature of about 25 °C (77 °F), are much more common. Cold water aquaria are for fish that are better suited to a cooler environment. More important than the range is consistency; most organisms are not accustomed to sudden changes in temperatures, which can cause shock and lead to disease. Water temperature can be regulated with a thermostat and heater (or cooler).

Water movement can also be important in simulating a natural ecosystem. Aquarists may prefer anything from still water up to swift currents, depending on the aquarium's inhabitants. Water movement can be controlled via aeration from air pumps, powerheads, and careful design of internal water flow (such as location of filtration system points of inflow and outflow).

Nitrogen Cycle

Of primary concern to the aquarist is management of the waste produced by an aquarium's inhabitants. Fish, invertebrates, fungi, and some bacteria excrete nitrogen waste in the form of ammonia (which converts to ammonium, in acidic water) and must then pass through the nitrogen cycle. Ammonia is also produced through the decomposition of plant and animal matter, including fecal matter and other detritus. Nitrogen waste products become toxic to fish and other aquarium inhabitants at high concentrations.

The Process

A well-balanced tank contains organisms that are able to metabolise the waste products of other aquarium residents. This process is known in the aquarium hobby as the nitrogen cycle. Bacteria known as nitrifiers (genus *Nitrosomonas*) metabolise nitrogen waste. Nitrifying bacteria capture ammonia from the water and metabolise it to producenitrite. Nitrite is toxic to fish in high concentrations. Another type of bacteria (genus *Nitrospira*) converts nitrite into nitrate, a less toxic substance. (*Nitrobacter*bacteria were previously believed to fill this role. While biologically they could theoretically fill the same niche as *Nitrospira*, it has recently been found that *Nitrobacter* are not present in detectable levels in established aquaria, while *Nitrospira* are plentiful.) However, commercial products sold as kits to "jump start" the nitrogen cycle often still contain *Nitrobacter*.

In addition to bacteria, aquatic plants also eliminate nitrogen waste by metabolising ammonia and nitrate. When plants metabolise nitrogen compounds, they remove nitrogen from the water by using it to build biomass that decays more slowly than ammonia-driven plankton already dissolved in the water.

Maintaining the Nitrogen Cycle

What hobbyists call the nitrogen cycle is only a portion of the complete cycle: nitrogen must be added to the system (usually through food provided to the tank inhabitants), and nitrates accumulate in the water at the end of the process, or become bound in the biomass of plants. The aquarium keeper must remove water once nitrate concentrations grow, or remove plants which have grown from the nitrates.

Hobbyist aquaria often do not have sufficient bacteria populations to adequately denitrify waste. This problem is most often addressed through two filtration solutions: Activated carbon filters absorb nitrogen compounds and other toxins, while biological filters provide a medium designed to enhance bacterial colonisation. Activated carbon and other substances, such as ammonia absorbing resins, stop working when their pores fill, so these components have to be replaced regularly.

New aquaria often have problems associated with the nitrogen cycle due to insufficient beneficial bacteria. Therefore fresh water has to be matured before stocking them with fish. There are three basic approaches to this: the "fishless cycle", the "silent cycle" and "slow growth".

In a fishless cycle, small amounts of ammonia are added to an unpopulated tank to feed the bacteria. During this process, ammonia, nitrite, and nitrate levels are tested to monitor progress. The "silent" cycle is basically nothing more than densely stocking the aquarium with fast-growing aquatic plants and relying on them to consume the nitrogen, allowing the necessary bacterial populations time to develop. According to anecdotal reports, the plants can consume nitrogenous waste so efficiently that ammonia and nitrite level spikes seen in more traditional cycling methods are greatly reduced or disappear. "Slow growth" entails slowly increasing the population of fish over a period of 6 to 8 weeks, giving bacteria colonies time to grow and stabilise with the increase in fish waste. This method is usually done with a small starter population of hardier fish which can survive the ammonia and nitrite spikes, whether they are intended to be permanent residents or to be traded out later for the desired occupants.

The largest bacterial populations are found in the filter, where is high water flow and plentiful surface available for their growth, so effective and efficient filtration is vital. Sometimes, a vigorous cleaning of the filter is enough to seriously disturb the biological balance of an aquarium. Therefore, it is recommended to rinse mechanical filters in an outside bucket of aquarium water to dislodge organic materials that contribute to nitrate problems, while preserving bacteria populations. Another safe practice consists of cleaning only half of the filter media during each service, or using two filters, only one of which is cleaned at a time.

Biological Loading

Biological load is a measure of the burden placed on the aquarium ecosystem by its inhabitants. High biological loading presents a more complicated tank ecology, which in turn means that equilibrium is easier to upset. Several fundamental constraints on biological loading depend on aquarium size. The water's surface area limits oxygen intake. The bacteria population depends on the physical space they have available to colonise. Physically, only a limited size and number of plants and animals can fit into an aquarium while still providing room for movement. Biologically, biological loading refers to the rate of biological decay in proportion to tank volume. Adding plants to an aquarium will sometimes help greatly with taking up fish waste as plant nutrients. Although an aquarium can be overloaded with fish, an excess of plants is unlikely to cause harm.

Calculating Capacity

Limiting factors include the oxygen availability and filtration processing. Aquarists have rules of humb to estimate the number of fish that can be kept in an aquarium. The examples below are for small freshwater fish; larger freshwater fishes and most marine fishes need much more generous allowances.

- 3 cm of *adult* fish length per 4 litres of water (i.e., a 6 cm-long fish would need about 8 litres of water).
- 1 cm of *adult* fish length per 30 square centimetres of surface area.
- 1 inch of *adult* fish length per US gallon of water.
- 1 inch of *adult* fish length per 12 square inches of surface area.

Experienced aquarists warn against applying these rules too strictly because they do not consider other important issues such as growth rate, activity level, social behaviour, filtration capacity, total biomass of plant life, and so on. Establishing maximum capacity is often a matter of slowly adding fish and monitoring water quality over time, following a trial and error approach.

Other Factors Affecting Capacity

One variable is differences between fish. Smaller fish consume more oxygen per gram of body weight than larger fish. Labyrinth fish can breathe atmospheric oxygen and do not need as much surface area (however, some of these fish are territorial, and do not appreciate crowding). Barbs also require more surface area than tetras of comparable size. Oxygen exchange at the surface is an important constraint, and thus the surface area of the aquarium matters. Some aquarists claim that a deeper aquarium holds no more fish than a shallower aquarium with the same surface area. The capacity can be improved by surface movement and water circulation such as through aeration, which not only improves oxygen exchange, but also waste decomposition rates.

Waste density is another variable. Decomposition in solution consumes oxygen. Oxygen dissolves less readily in warmer water; this is a double-edged sword since warmer temperatures make fish more active, so they consume more oxygen.

In addition to bioload/chemical considerations, aquarists also consider the mutual compatibility of the fish. For instance, predatory fish are usually not kept with small, passive species, and territorial fish are often unsuitable tankmates for shoaling species. Furthermore, fish tend to fare better if given tanks conducive to their size. That is, large fish need large tanks and small fish can do well in smaller tanks. Lastly, the tank can become overcrowded without being overstocked. In other words, the aquarium can be suitable with regard to filtration capacity, oxygen load, and water, yet still be so crowded that the inhabitants are uncomfortable.

Aquarium Classifications

From the outdoor ponds and glass jars of antiquity, modern aquaria have evolved into a wide range of specialised systems. Individual aquaria can vary in size from a small bowl large enough for only a single small fish, to the huge public aquaria that can simulate entire marine ecosystems.

One way to classify aquaria is by salinity. Freshwater aquaria are the most popular due to their lower cost. More expensive and complex equipment is required to set up and maintain a marine aquaria. Marine aquaria frequently feature a diverse range of invertebrates in addition to species of fish. Brackish water aquaria combine elements of both marine and freshwater fish keeping. Fish kept in brackish water aquaria generally come from habitats with varying salinity, such as mangrove swamps and estuaries. Subtypes exist within these types, such as the reef aquarium, a typically smaller marine aquarium that houses coral.

Another classification is by temperature range. Many aquarists choose a tropical aquarium because tropical fish tend to be more colourful. However, the coldwater aquariumis also popular, which is mainly restricted to goldfish, but can include fish from temperate areas worldwide and native fish keeping.

Aquaria may be grouped by their species selection. The community tank is the most common today, where several non-aggressive species live peacefully. In these aquaria, the fish, invertebrates, and plants probably do not originate from the same geographic region, but tolerate similar water conditions. Aggressive tanks, in contrast, house a limited number of species that can be aggressive toward other fish, or are able to withstand aggression well. Most marine tanks and tanks housing cichlids have to take the aggressiveness of the desired species into account when stocking. Specimen tanks usually only house one fish species, along with plants, perhaps ones found in the fishes' natural environment and decorations simulating a natural ecosystem. This type is useful for fish that cannot coexist with other fish, such as the electric eel, as an extreme example. Some tanks of this sort are used simply to house adults for breeding.

Ecotype, ecotope, or biotope aquaria is another type based on species selection. In it, an aquarist attempts to simulate a specific natural ecosystem, assembling fish, invertebrate species, plants, decorations and water conditions all found in that ecosystem. These biotope aquaria are the most sophisticated hobby aquaria; public aquaria use this approach whenever possible. This approach best simulates the experience of observing in the wild. It typically serves as the healthiest possible artificial environment for the tank's occupants.

Public Aquaria

Most public aquarium facilities feature a number of smaller aquaria, as well those too large for home aquarists. The largest tanks hold millions of gallons of water and can house large species, including sharks or beluga whales. Dolphinaria are specifically for dolphins. Aquatic and semiaquatic animals, including otters and penguins, may also be kept by public aquaria. Public aquaria may also be included in larger establishments such as a marine mammal park or a marine park.

Virtual Aquariums

A virtual aquarium is a computer program which uses 3D graphics to reproduce an aquarium on a personal computer. The swimming fish are rendered in real time, while the background of the tank is usually static. Objects on the floor of the tank may be mapped in simple planes so that the fish may appear to swim both in front and behind them, but a relatively simple 3D map of the general shape of such objects may be used to allow the light and ripples on the surface of the water to cast realistic shadows.

Bubbles and water noises are common for virtual aquariums, which are often used as screensavers.

The number of each type of fish can usually be selected, often including other animals like starfish, jellyfish, seahorses, and even sea turtles. Other objects found in an aquarium can also be added and rearranged on some software, like treasure chests and giant clams that open and close with air bubbles, or a bobbing diver. There are also usually features that allow the user to tap on the glass or put food in the top, both of which the fish will react to. Some also have the ability to allow the user to fish and other objects to create new varieties.

Aquascaping

Aquascaping is the craft of arranging aquatic plants, as well as rocks, stones, cavework, or driftwood, in an aesthetically pleasing manner within an aquarium—in effect, gardening under water. Aquascape designs include a number of distinct styles, including the garden-like Dutch style and the Japanese-inspired nature style. Typically, an aquascape houses fish as well as plants, although it is possible to create an aquascape with plants only, or with rockwork or other hardscape and no plants.

Although the primary aim of aquascaping is to create an artful underwater landscape, the technical aspects of aquatic plant maintenance must also be taken into consideration. Many factors must be balanced in the closed system of an aquarium tank to ensure the success of an aquascape. These factors include filtration, maintaining carbon dioxide at levels sufficient to support photosynthesis underwater, substrate and fertilization, lighting, and algae control.

Aquascape hobbyists trade plants, conduct contests, and share photographs and information via the internet. The United States-based Aquatic Gardeners Association has about 1,200 members.

Designs

Dutch Style

The Dutch aquarium employs a lush arrangement in which multiple types of plants having diverse leaf colours, sizes, and textures are displayed much as terrestrial plants are shown in a flower garden. This style was developed in the Netherlands starting in the 1930s, as freshwater aquarium equipment became commercially available. It emphasizes plants located on terraces of different heights, and frequently omits rocks and driftwood.

Linear rows of plants running left-to-right are referred to as "Dutch streets". Although many plant types are used, one typically sees neatly trimmed groupings of plants with fine, feathery foliage, such as *Limnophila aquatica* and various types of *Hygrophila*, along with the use of red-leaved *Alternanthera reineckii*, *Ammania gracilis*, and assorted *Rotala* for colour highlights. More than 80% of the aquarium floor is covered with plants, and little or no substrate is left visible. Tall growing plants that cover the back glass originally served the purpose of hiding bulky equipment behind the tank.

Nature Style

A contrasting approach is the "nature aquarium" or Japanese style, introduced in the 1990s by Takashi Amano. Amano's three-volume series, *Nature Aquarium World,* sparked a wave of interest in aquarium gardening, and he has been cited as having "set a new standard in aquarium management". Amano's compositions draw on Japanese gardening techniques that attempt to mimic natural landscapes by the asymmetrical arrangement of masses of relatively few species of plants, and carefully selected stones or driftwood. The objective is to evoke a landscape in miniature, rather than a colourful garden.

This style draws particularly from the Japanese aesthetic concepts of *Wabi-sabi*, which focuses on transience and minimalism as sources of beauty, and *Iwagumi*, which sets rules governing rock placement. In the *Iwagumi* system, the *Oyaishi* or main stone, is placed slightly off-centre in the tank, and *Soeishi* or accompanying stones, are grouped near it, while *Fukuseki*, or secondary stones, are arranged in subordinate positions. The location of the focal point of the display, determined largely by the asymmetric placement of the *Oyaishi*, is considered important, and follows ratios that reflect Pythagorean tuning. Plants with small leaves, such as *Eleocharis acicularis*, *Glossostigma elatinoides*, *Hemianthus callitrichoides*, *Riccia fluitans*, small aquatic ferns, and Java moss (*Versicularia dubyana*or *Taxiphyllum barbieri*) are usually emphasized, with more limited colours than in the Dutch style, and the hardscape is not completely covered. Fish, or freshwater shrimp such as *Caridina multidentata* and *Neocaridina heteropoda*, are usually selected to complement the plants and control algae.

Jungle Style

Some hobbyists also refer to a "jungle" (or "wild jungle") style, separate from either the Dutch or nature styles, and incorporating some of the features of them both. Bold, coarser leaf shapes, such as *Echinodorus bleheri*, are used to provide a wild, untamed appearance.

Biotopes

The styles above often combine plant and animal species based on the desired visual impact, without regard to geographic origin. Biotope aquascapes are designed instead to replicate exactly a particular aquatic habitat at a particular geographic location, and not necessarily to provide a garden-like display. Plants and fish need not be present at all, but if they are, they must match what would be found in nature in the habitat being represented, as must any gravel and hardscape, and even the chemical composition of the water.

Paludariums

In a paludarium, part of the aquarium is underwater, and part is above water. Substrate is built up so that some "land" regions are raised above the waterline, and the tank is filled with water only part way. This allows plants, such as *Cyperus alternifolius* and *Spathiphyllum wallisii*, as well as various *Anubias* and some bromeliads, to grow emersed, with their roots underwater but their tops in the air, as well as completely submersed. In some configurations, plants that float on the surface of the water, such as *Eichhornia crassipes* and *Pistia stratiotes*, can be displayed to full advantage. Unlike other aquarium setups, paludariums are particularly well-suited to keeping amphibians.

Saltwater Reefs

Dutch and nature style aquascapes are traditionally freshwater systems. In contrast, relatively few ornamental plants can be grown in a saltwater aquarium. Saltwater aquascaping typically centres, instead, on mimicking a reef. An arrangement of live rock forms the main structure of this aquascape, and it is populated by corals and other marine invertebrates as well as coralline algae, which together serve much the same aesthetic role as freshwater plants.

Lighting plays a particularly significant role in the reef aquascape. Many corals, as well as tridacnid clams, contain symbiotic fluorescent algae-likeprotozoa called zooxanthellae. By providing intense lighting supplemented in the ultraviolet wavelengths, reef aquarists not only support the health of these invertebrates, but also elicit particularly bright colours emitted by the fluorescent microorganisms.

Techniques

In addition to design, freshwater aquascaping also requires specific methods to maintain healthy plants underwater. Plants are often trimmed to obtain the desired shape, and they can be positioned by tying them in place inconspicuously with thread. Most serious aquascapers use aquarium-safe fertilizers, commonly in liquid or tablet form, to help the plants fill out more rapidly. Some aquarium substrates containing laterite also provide nutrients.

It is also necessary to support photosynthesis, by providing light and carbon dioxide. A variety of lighting systems may be used to produce the full spectrum of light, usually at 2–4watts per gallon (0.5–1 watts per litre). Lights are usually controlled by a timer that allows the plants to be acclimated to a set cycle. Depending on the number of plants and fish, the aquascape may also require carbon dioxide supplementation. This can be accomplished with a simple homemade system, using a soda bottle filled with yeast, warm water, and sugar, and connected to an airstone in the aquarium, or with a pressurized CO_2 tank that injects a set amount of carbon dioxide into the aquarium water.

Algae (including cyanobacteria, as well as true algae) is considered distracting and unwanted in aquascaping, and is controlled in several ways. One is the use of animals that consume algae, such as some fish (notably cyprinids of the genera *Crossocheilus* and *Gyrinocheilus*, and catfish of the genera *Ancistrus*, *Hypostomus*, and *Otocinclus*), shrimp, orsnails, to clean the algae that collects on the leaves. A second is using adequate light and CO_2 to promote rapid growth of desired plants, while controlling nutrient levels, to ensure that the plants utilise all fertilizer without leaving nutrients to support algae.

Although serious aquascapers often use a considerable amount of equipment to provide lighting, filtration, and CO_2 supplementation to the tank, some hobbyists choose instead to maintain plants with a minimum of technology, and some have reported success in producing lush plant growth this way. This approach, sometimes called the "natural planted tank" and popularised by Diana Walstad, can include the use of soil in place of aquarium gravel, the elimination of CO_2 apparatus and most filtration, and limited lighting. Instead, only a few fish are kept, to limit the quantity of fish waste, and the plants themselves are used to perform the water-cleansing role typically played by aquarium filters, by utilising what fish waste there is as fertilizer.

Contests

Early Dutch hobbyists began the practice of aquascape contests, with over 100 local clubs. Judges had to go through about three years of training and pass examinations in multiple disciplines in order to qualify. This competition continues to be held every year, under the auspices of the National Aquarium Society. There are three rounds, beginning with contests in local clubs. First-place local winners are entered in the second round, held in fifteen *districtkeuring* (districts). The winners at that level are then entered in the third round, which is the national championship.

In the Dutch contest, the focus is not only on composition, but also on the biological well-being of the aquarium's inhabitants. Most points are, in fact, awarded for such biological criteria as fish health, plant health, and water quality. Unlike contests in other countries, the judges travel to each contestant's home to evaluate the tank, where they measure the water parameters themselves.

The Aquatic Gardeners Association, based in the United States, Aqua Design Amano, based in Japan, and Aquatic Scapers Europe, based in Germany, also conduct annual freshwater aquascaping contests. Entries from around the world are submitted as photographs and explanatory text online.

The Aquatic Gardeners Association contest is judged based on:

- overall impression (35 points),
- composition, balance, use of space and use of colour (30 points),
- selection and use of materials (20 points), and
- viability of aquascape (15 points).

There are also smaller contests conducted by Acuavida in Spain, and by the Greek Aquarist's Club.

Public Aquariums

Large public aquariums sometimes use aquascaping as part of their displays. As early as the 1920s, the New York Aquarium included a moray eel display tank that was decorated with calcareous tufa rock, arranged to resemble a coral reef, and supporting some stony corals and sea fans. Because they typically present wildlife from a particular habitat, modern day displays are often created to be biologically accurate biotopes.

Association of Zoos and Aquariums

The Association of Zoos and Aquariums (previously American Zoo and Aquarium Association, and originally American Association of Zoological Parks and Aquariums) was founded in 1924 and is a nonprofit organisation dedicated to the advancement of zoos and public aquariums in the areas of conservation, education, science, and recreation. The AZA headquarters is located in Silver Spring, Maryland.

History

In October 1924 the American Association of Zoological Parks and Aquariums (AAZPA) was formed as an affiliate of the American Institute of Park Executives (AIPE). In 1966, the AAZPA became a professional branch affiliate of the newly formed National Recreation and Park Association (NRPA which absorbed the AIPE). In the fall of 1971, the AAZPA membership voted to become an independent association and, in January 1972, it was chartered as the AAZPA with its executive office located in Wheeling, West Virginia within the Oglebay Park Good Zoo. In

January 1994, the shorter name American Zoo and Aquarium Association (AZA) was adopted.

In 1998, there were 134 million visits to North American zoos and aquariums. Ten years later, in 2008, there were 175 million visits to AZA-accred zoos and aquariums.

Activities

The organisation is active in institution accration, animal care initiatives, education and conservation programs, collaborative research and political lobbying in order to achieve this goal. It serves as an accring body for zoos and aquariums and ensures that accred facilities meet higher standards of animal care than are required by law. Institutions are evaluated every five years in order to ensure standards are met and to maintain accration. The association also facilitates both Species Survival Plans and Population Management Plans, which serve to sustainably manage genetically diverse captive populations of various animal species. In addition, the association sponsors the scientific journal *Zoo Biology*.

Annual Report on Conservation and Science

The association has established a computerised database called the *Annual Report on Conservation and Science*. It provides a model for a broader database to help track research projects worldwide. The database can be searched by key word, name of researcher, topic, country or region, name of institution, conservation program title, name of cooperating institution (including governmental agencies and non-governmental organisations, colleges or universities, and non-member zoos and aquariums), type of research, or date.

In 2000-2001, member institutions reported that they participated in over 2,230 conservation projects (1,390 in situ and 610 ex situ, 230 both) in 94 countries. They published 1,450 books, book chapters, journal articles, conference proceeding papers, posters and theses or dissertations. The publications can be searched using keywords, name of author, type of publication, institution name, or date.

Regulation

In the United States, any public animal exhibit must be licensed and inspected by the United States Department of Agriculture (USDA), United States Environmental Protection Agency, Drug Enforcement Agency, Occupational Safety and Health Administration, and others. Depending on the animals they exhibit, the activities of zoos are regulated by laws including the Endangered Species Act, the Animal Welfare Act, the Migratory Bird Treaty Act of 1918 and others. Additionally, zoos in North America may choose to pursue accration by the Association of Zoos and Aquariums (AZA). The American association has developed a definition for zoological gardens and aquariums as part of its accration standards: "A permanent cultural institution which owns and maintains captive wild animals that represent more than a token collection and, under the direction of a professional staff, provides its collection with appropriate care and exhibits them in an aesthetic manner to the public on a regularly scheduled basis. They shall further be defined as having as their primary business the exhibition, conservation and preservation of the earth's fauna in an educational and scientific manner." To achieve accration, a zoo must pass an application and inspection process and meet or exceed the AZA's standards for animal health and welfare, fundraising, zoo staffing, and involvement in global conservation efforts. Inspection is performed by three experts (typically one veterinarian, one expert in animal care, and one expert in zoo management and operations) and then reviewed by a panel of twelve experts before accration is awarded. This accration process is repeated once every five years. The AZA estimates that there are approximately 2,400 animal exhibits operating under USDA license as of February 2007; fewer than 10% are accred.

Filter (Aquarium)

Aquarium filters are critical components of both freshwater and marine aquaria. Aquarium filters remove physical and soluble chemical waste products from aquaria, simplifying maintenance. Furthermore, aquarium filters are necessary to support life as aquaria are relatively small, closed volumes of water compared to the natural environment of most fish.

Overview

Animals, typically fish, kept in fish tanks produce waste from excrement and respiration. Another source of waste is uneaten food or plants and fish which have died. These waste products collect in the tanks and contaminate the water. As the degree of contamination rises, the risk to the health of the aquaria increases and removal of the contamination becomes critical. Filtration is a common method used for maintenance of healthy aquaria.

Biological Filtration and the Nitrogen Cycle

Proper management of the nitrogen cycle is a vital element of a successful aquarium. Excretia and other decomposing organic matter produce ammonia which is highly toxic to fish. Bacterial processes oxidise this ammonia into the slightly less toxic nitrites, and these are in turn oxidised to form the much less toxic nitrates. In the natural environment these nitrates are subsequently taken up by plants as fertilizer and this does indeed happen to some extent in an aquarium planted with real plants.

An aquarium is, however, an imperfect microcosm of the natural world. Aquariums are usually much more densely stocked with fish than the natural environment. This increases the amount of ammonia produced in the relatively small volume of the aquarium. The bacteria responsible for breaking down the ammonia colonise the surface of any objects inside the aquarium. In most case, a biological filter is nothing more than a chemically inert porous sponge, which provides a greatly enlarged surface area on which these bacteria can develop. These bacterial colonies take several weeks to form, during which time the aquarium is vulnerable to a condition commonly known as "new tank syndrome" if stocked with fish too quickly. Some systems incorporate bacteria capable of converting nitrates into nitrogen gas.

Accumulation of toxic ammonia from decomposing wastes is the largest cause of fish mortality in new, poorly maintained, or overloaded aquariums. In the artificial environment of the aquarium, the nitrogen cycle effectively ends with the production of nitrates. In order that the nitrate level does not build up to a harmful level regular partial water changes are required to remove the nitrates and introduce new, uncontaminated water.

Mechanical and Chemical Filtration

The process of mechanical filtration removes particulate material from the water column. This particulate matter may include uneaten food, feces or plant or algal debris. Mechanical filtration is typically achieved by passing water through materials which act as a sieve, physically trapping the particulate matter. Removal of solid waste can be as simple as physical hand netting of debris, and/or involve highly complex equipment. All removal of solid wastes involve filtering water through some form of mesh in a process known as mechanical filtration. The solid wastes are first collected, and then must be physically removed from the aquarium system. Mechanical filtration is ultimately ineffective if the solid wastes are not removed from the filter, and are allowed to decay and dissolve in the water.

Dissolved wastes are more difficult to remove from the water. Several techniques, collectively known as chemical filtration, are used for the removal of dissolved wastes, the most popular being the use of activated carbon and foam fractionation. To a certain extent, healthy plants extract dissolved chemical wastes from water when they grow, so plants can serve a role in the containment of dissolved wastes.

A final and less common situation requiring filtration involves the desire to sterilize water born pathogens. This sterilization is accomplished by passing aquarium water through filtration devices which expose the water to high intensity ultraviolet light and/or exposing the water to dissolved ozone gas.

Materials Suitable for Aquarium Filtration

Numerous materials are suitable as aquarium filtration media. These include synthetic wools, known in the aquarium hobby as filter wool, made of polyethylene terephthalate or nylon.Synthetic sponges or foams, various ceramic and sintered glass and silicon products along with igneous gravels are also used as mechanical filter materials. Materials with a greater surface area provide both mechanical and biological filtration. Some filter materials, such as plastic "bioballs", are best used for biological filtration.

With the notable exception of diatom filters, aquarium filters are rarely purely mechanical in action, as bacteria will colonise most filter materials effecting some degree of biological filtration. Activated carbon and zeolites are also frequently added to aquarium filters. These highly porous materials act as adsorbates binding various chemicals to their large external surfaces and also as sites of bacterial colonisation.

The simplest type of aquarium filter consists only of filter wool and activated carbon. The filter wool traps large debris and particles, and the activated carbon adsorbs smaller impurities. These should be changed regularly at suitable intervals. This is particularly important in the case of activated carbon filters, which may re-release their adsorbed contents in large (and therefore harmful) doses if they are allowed to saturate. Activated carbon adsorbs toxins on the surface of the carbon. It can be reactivated by boiling in clean water, and draining away the tainted water, and can then be re-used.

Types of Aquarium Filters

Numerous types of aquarium filters are commercially available, including:

External Filters

External filters remove water from the aquarium which is then pushed (or pulled) through a series of different levels of filter media and returned to the aquarium. They are usually more effective and easier to maintain than internal filters.

Canister Filters

Compared to filters that hang on the back of the aquarium, canister-style external filters offer a greater quantity of filter materials to be used along with a greater degree of flexibility with respect to filter material choice. Water enters the canister filled with the chosen filter material through an intake pipe at the bottom of the canister, passes through the material, and is fed back to the aquarium through the return pipe. Water is forced to circulate through the filter by a pump typically installed at the top of the canister. It is important to note that canister filters are sealed, fully flooded systems, meaning that the aquarium, intake pipe, filter interior and the return pipe form a continuous body of water. In this configuration both the intake and return path form two siphons, which precisely counterbalance each other. Under these circumstances, the filter pump does not have to spend any effort to lift the water back to the aquarium, regardless of how high the latter is installed above the canister. The pump should only be powerful enough to push the water through the filtering material as well as overcome the drag in the intake and return pipes. This makes canister filter pumps virtually insensitive to the height difference between the aquarium and the filter (although exceeding the manufacturer-specified height limit can lead to leaks).

Benefits of this type of filter are that they can provide a high volume of filter material without reducing the internal space in the aquarium, and that they can be disconnected from the tank for cleaning/maintenance and replaced without disturbing the aquarium interior or occupants. Also, as a filter with external plumbing, it supports in-line installation of other aquarium equipment, such as water heaters and carbon dioxide diffusers. Such equipment can be removed from the tank and installed in-line into the return pipe of the filter. Disadvantages of canister filters include the increased cost and complexity relative to internal filters and difficulties in cleaning the tubes which transfer water to and from the aquarium. There is also the risk of a leak, which naturally is an issue for any filter placed outside of the aquarium.

Canister filters were initially designed to filter drinking water, under low pressure. Canister filters for aquariums use high water pressure, from a properly powered pump, to force water through the dense filter media. A pump can draw water from an under-gravel filter, and run it into a canister for double filtration.

Diatom Filters

Diatom filters are used only for sporadic cleaning of tanks, they are not continuously operated on aquariums. These filters utilise diatomaceous earth to create an extremely fine filter down to 1 μm which removes particulate matter from the water column.

Trickle Filters

Trickle filters, also known as wet/dry filters are another water filtration systems for marine and freshwater aquariums. This filter comes in two configurations, one which is placed on top of the aquarium (more rarely seen) and one which is placed below the aquarium (more common).

If the wet/dry filter is placed on top of the aquarium, water is pumped over a number of perforated trays containing filter wool or some other filter material. The water trickles through the trays, keeping the filter wool wet but not completely submerged, allowing aerobic bacteria to grow and aiding biological filtration. The water returns to the aquarium like rain.

Alternatively, the wet/dry filter may be placed below the tank. In this design, water is fed by gravity to the filter below the aquarium. Prefiltered water is delivered to a perforated plate (drip plate). Prefiltering may take place in the aquarium via a foam block or sleeve in the overflow, or weir siphon, or it may be prefiltered by filter wool resting on the perforated plate. The waste laden water from the aquarium spreads over the drip plate, and rains down through a medium.

This may be a filter wool/plastic grid rolled into a circular shape (DLS or "Double Layer Spiral") or any number of plastic media commonly known as Bio Balls. As the water cascades over the media, CO_2 is given off, oxygen is picked up, and bacteria convert the waste from the tank into less harmful materials. From here the water enters the sump. The sump may contain a number of compartments, each with its own filtration material. Often, heaters and thermostats are placed in the sump.

Fluidized Bed Filter

The fluidized bed filter (FBF) is a biological reactor only. The principle is to direct water through a sand (or similar media) bed from below so that the sand becomes fluidized – behaves like a fluid. This mechanism is seen in liquifaction, quick sand, and industrial processes including municipal sewage treatment. The combined surface of all sand particles in the filter is very large, and so there is a large surface for aerobic denitrification bacteria. Therefore the size of the filter can be modest. The filter itself can be internal or external. In its simplest DIY internal version an FBF is very easy to build, with a container, sand, pump, and some plumbing. There are many variables: shape and size of the container, quantity of sand or equivalent, particle sizes, the pump's power, and plumbing.

Algae Filters

Algae may be grown purposely, which removes chemicals from the water which need to be removed in order to have healthy fish, invertebrates and corals. This is a natural ("green") filtering method, which lets an aquarium operate the way oceans and lakes operate.

Baffle Filters

Baffle filters are similar to wet and dry, trickle filters in that they are generally situated below the aquarium. This type of filter consists of a series of baffles that the water must pass through in order to reach the pump which is returning water to the aquarium. These baffles then act much like a series of canister filters and can be filled with different filter media for different purposes.

Internal Filters

Internal filters are, by definition, filters within the confines of the aquarium. These include the sponge filter, variations on the corner filter (pictured top right and left), foam cartridge filter and the undergravel filter. An internal filter may have an electric pump and thus be an internal power filter, often attached to the inside of aquaria via suction cups.

Airlift Filters

Sponge filters and corner filters (sometimes called box filters) work by essentially the same mechanism as an internal filter. Both generally work by airlift, using bubbles from an air pump rising in a tube to create flow. In a sponge filter, the inlet may only be covered by a simple open-cell block of foam. A corner filter is slightly more complex. These filters are often placed in the corner on the bottom of the aquarium. Water enters slits in the box, passes through a layer of medium, then exits through the airlift tube to return to the aquarium. These filters tend to only be suitable for small and lightly stocked aquaria. The sponge filter is especially useful for rearing fry where the sponge prevents the small fish from entering the filter.

Undergravel Filters

One of the oldest types of filters, the undergravel filters consist of a porous plate which is placed beneath the gravel on the base of the aquarium and one, or more, uplift tubes. Historically, undergravel filters have been driven via air displacement. Air stones are placed at the base of uplift tubes which force water out of the uplift tube creating negative pressure beneath the undergravel filter plate. Water then percolates down through the gravel which itself is the filtration material. Greater flow rate of water through the gravel can be achieved via the use of water pump rather than air displacement.

Beneficial bacteria colonise the gravel bed and provide biological filtration, using the substrate of the aquarium itself as a biological filter.

Undergravel filters can be detrimental to the health of aquatic plants. Fine substrates such as sand or peat may clog an undergravel filter. Undergravel filters are not effective if the substrate bed is uneven. In an uneven gravel bed, water will flow only through the thin portions of the bed, leaving the more heavily covered areas to become anoxic. Because of this, animals that dig, such as cichlids, are best kept in an aquarium using some other type of filtration.

Aquarium Furniture

Aquarium furniture refers to the various ornaments and functional items in an aquarium.

Ornamental aquarium furniture is often kitsch: popular examples include ceramic mermaids, 'sunken' ships and castles, and the ever-popular (but curiously misplaced) "No Fishing" sign. Another strange piece of decor is the ubiquitous plastic corals found often in freshwater tanks.

Examples of functional aquarium furniture would include devices for removing algae from the glass (either a razor or a scouring pad, attached to the glass by a magnet), airstones, water filters, water heaters, and food dispensers.

Aquarium furniture may also refer to an item of (regular) furniture that features an aquarium in its design. A stand or cabinet that supports the aquarium may be considered aquarium furniture. Also, many home reef aquariums canopies containing metal halide lights. The canopies are often constructed to the same standards as high quality cabinetry.

Live Rock

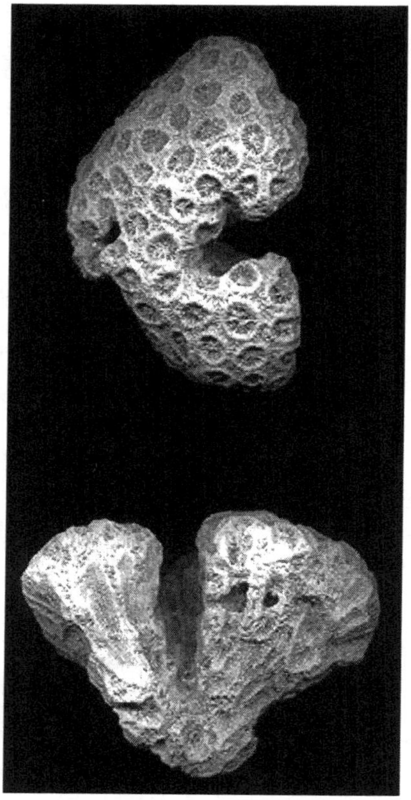

Figure: Bleached coral skeletons, which can be inhabited by micro- and macro-organisms to form live rock

Live rock is rock from the ocean that has been introduced into a saltwater aquarium. Along with live sand, it confers to the closed marine system multiple benefits desired by the saltwater aquarium hobbyist. The name sometimes leads to misunderstandings, as the "live rock" itself is not actually alive, but is made simply from the aragonite skeletons of long dead corals, or other calcareous organisms, which in the ocean form the majority of coral reefs. When taken from the ocean it is usually encrusted with coralline algaeand inhabited by a multitude of marine organisms. The many forms of micro and macroscopic marine life that live on and inside of the rock, which acts as an ideal habitat, give it the name "live rock".

Origin

Live rock is harvested for use in the aquarium trade from collections in the wild near reefs, where parts may become detached from the main body of coral by storms. It may also be "seeded" from small coralline rocks by an aquaculturalist in warm ocean water, to be harvested later. Live rock can also be seeded by adding base rock to an active reef aquarium that already has live rock. Live rock harbours a wide variety of corals, algae, sponges, and other invertebrates, when they are collected. Corals added to the aquarium later will often become attached to the rock.

Purpose

For the aquarium trade this rock is highly valued not only for the diversity of life it can bring to the closed marine environment, but its function as a superior biological filter that hosts both aerobic and anaerobic nitrifying bacteria required for the nitrogen cycle that processes waste. Live rock becomes the main biological nitrification base or biological filter of a saltwater aquarium. Additionally, live rocks have a stabilising effect on the water chemistry, in particular on helping to maintain constant pH by release of calcium carbonate. Lastly, live rock, especially when encrusted with multiple colours of coralline algae, becomes a major decorative element of the aquarium and provides shelter for the inhabitants. It is often used to build caves, arches, overhangs, or other structures in the tank, a practice known as aquascaping.

In J. Charles Delbeek's article *Your First Reef Aquarium*, he states,

> *The use of live rock immediately introduces into the aquarium numerous algae, bacteria and small invertebrates all of which contribute to the overall quality of the aquarium water. Live rock has just as much, if not more, surface area for bacteria than a trickle filter. Since live rock in the aquarium contains various types of bacteria, algae and corals, waste products such as ammonia, nitrate and phosphate can have a number of fates. Ammonia, nitrate and phosphate are readily assimilated by algae and photosynthetic corals growing on and in the rock. Ammonia can also be quickly converted into nitrate by the bacteria on and in the rock. This nitrate can be either absorbed by the algae and corals, or it can be denitrified by bacteria in close proximity to the nitrate-producing bacteria.*

Types

There are many different types of live rock. Each is named after the area from which it originated. A large amount of live rock comes from the Southern Pacific region, in areas such as Fiji, Tonga, and the Marshall Islands, as well as from the Caribbean. Each has its own distinct qualities that make it preferable to certain reef aquarists. For instance, live rock from the Fiji region is often porous and large, and rock from the Tonga region is often dense and elongated.

Base Rock

Base rock is a generic term for aragonite rock that has no organisms growing in or on the rock. Base rock is often used as filler rock in the aquarium as it is much cheaper to purchase than live rock. In time, base rock will become colonised by living organisms.

Recently base rock that is mined from inland ancient reefs has become a popular way to keep the aquarium trade going sustainably. This rock is either maricultured and sold as live rock, or can be purchased and grown in the home aquarium.

Base rock can also be made from artificial rock called aragocrete, which is a hand made concrete from combining crushed aragonite, sand, and Portland cement. After allowing the cement to dry, the pieces are sometimes acid washed to counteract the high pH of the materials, and then allowed to soak in clean water for one or more months. They generally tend to be heavier and less attractive when compared to natural base rock.

Collection Ban

As of August 4, 2008 CITES (Convention on International Trade in Endangered Species) banned the collection of live rock from Tonga, the Marshall Islands, and the Cook Islands. This is due to the over-collecting of rock in these areas.

The Secretariat has lifted all of the CITES restrictions in the recently banned countries for exporting species listed in Appendix II. This includes Tonga, Fiji, and Australia. Most notably, Tonga rock and corals and Fiji corals are available once again.

Coral

Corals are marine animals in class Anthozoa of phylum Cnidaria typically living in compact colonies of many identical individual "polyps". The group includes the important reefbuilders that inhabit tropical oceans and secrete calcium carbonate to form a hard skeleton.

A coral "head" is a colony of myriad genetically identical polyps. Each polyp is a spineless animal typically only a few millimetres in diametre and a few centimetres in length. A set of tentacles surround a central mouth opening. An exoskeleton is excreted near the base. Over many generations, the colony thus creates a large skeleton that is characteristic of the species. Individual heads grow by asexual reproduction of polyps. Corals also breed sexually by spawning: polyps of the same species release gametes simultaneously over a period of one to several nights around a full moon.

Although corals can catch small fish and plankton, using stinging cells on their tentacles, most corals obtain the majority of their energy and nutrients from photosynthetic unicellular algae called zooxanthellae that live within the coral's tissue. Such corals require sunlight and grow in clear, shallow water, typically at depths shallower than 60 metres (200 ft). Corals can be major contributors to the physical structure of the coral reefs that develop in tropical and subtropical waters, such as the enormous Great Barrier Reef off the coast of Queensland, Australia. Other corals do not have associated algae and can live in much deeper water, with the cold-water genus *Lophelia* surviving as deep as 3,000 metres (9,800 ft). Examples live on the Darwin Mounds located north-west of Cape Wrath, Scotland. Corals have also been found off the coast of the U.S. in Washington State and the Aleutian Islands in Alaska.

Taxonomy

Corals divide into two subclasses, depending on the number of tentacles or lines of symmetry, and a series of orders corresponding to their exoskeleton, nematocyst type and mitochondrial genetic analysis. Common coral typing crosses suborder/class boundaries.

Hermatypic Corals

Hermatypic corals in the subclass Scleractinia are stony corals that build reefs. They mostly obtain at least part of their energy requirements from zooxanthellae, symbioticphotosynthetic microalgae. They secrete calcium carbonate to form a hard skeleton. Those having six or fewer lines of symmetry in their body structure are called hexacorallia orZoantharia. This group includes reef-building corals (scleractinians), sea anemones and zoanthids. Hermatypic genera include *Scleractinia*, *Millepora*, *Tubipora* and *Heliopora*.

In the Caribbean alone, at least 50 species of uniquely structured hard coral exist. Well-known types include:

- Brain corals grow to 1.8 metres (5.9 ft) in width.
- *Acropora* and staghorn corals grow fast and large, and are important reef-builders. Staghorn coral displays large, antler-like branches, and grows in areas with strong surf.
- Pillar coral forms pillars which can grow to 3 metres (9.8 ft) in height.
- *Leptopsommia*, or rock coral, appears almost everywhere in the Caribbean.

Ahermatypic Corals

Ahermatypic corals have no zooxanthellae. They sport eight tentacles and are also called octocorallia. They include corals in subclass *Alcyonacea*, as well as some species in order *Anthipatharia* (black coral, *Cirripathes*, *Antipathes*). Ahermatypic corals, such as sea whips, sea feathers, and sea pens, are also known as soft corals.

Unlike stony corals, they are flexible, undulating in the current, and often are perforated, with a lacy appearance. Their skeletons are proteinaceous, rather than calcareous. Soft corals are somewhat less plentiful (in the Caribbean, twenty species appear) than stony corals.

Perforate Corals

Corals can be perforate or imperforate. Perforate corals have porous skeletons, which allows their polyps to connect with each other through the skeleton. Imperforate corals have hard solid skeletons.

Anatomy

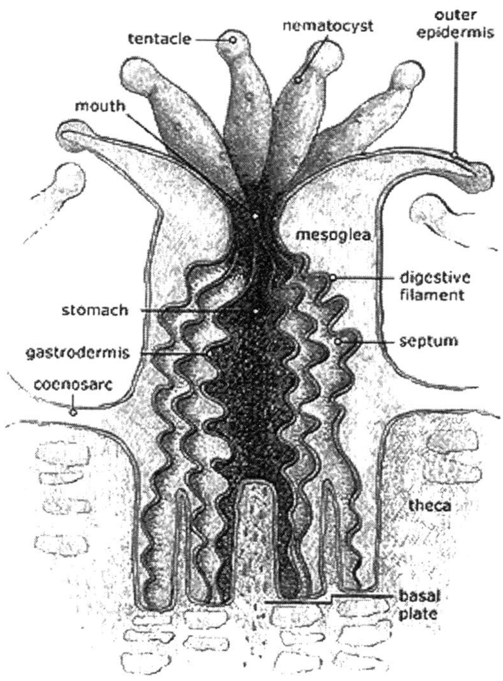

Figure: Anatomy of a coral polyp

Colonial Form: The polyps interconnect by a complex and well-developed system of gastrovascular canals, allowing significant sharing of nutrients and symbiotes. In soft corals, these range in size from 50–500 micrometres (0.0020–0.020 in) in diametre, and allow transport of both metabolites and cellular components.

Polyp: While the coral head is the familiar visual form of a single organism, it is actually a group of many individual, yetgenetically identical, multicellular organisms known as polyps. Polyps are usually a few millimetres in diametre, and are formed by a layer of outer epithelium and inner jellylike tissue known as the mesoglea. They are radially symmetrical, with tentacles surrounding a central mouth, the only opening to the stomach or coelenteron, through which food is ingested and waste expelled.

Exoskeleton: The stomach closes at the base of the polyp, where the epithelium produces an exoskeleton called the basal plate or calicle (L. small cup). The calicle is formed by a thickened calcareous ring (annular thickening) with six supporting radial ridges. These structures grow vertically and project into the base of the polyp. When a polyp is physically stressed, its tentacles contract into the calyx so that virtually no part is exposed above the skeletal platform. This protects the organism from predators and the elements.

The polyp grows by extension of vertical calices which occasionally septate to form a new, higher, basal plate. Over many generations, this extension forms the large calcareous structures of corals and ultimately coral reefs.

Formation of the calcareous exoskeleton involves deposition of the mineral aragonite by the polyps from calcium and carbonate ions they acquire from seawater. The rate of deposition, while varying greatly across species and environmental conditions, can reach 10 g/m^2 of polyp/day (0.3 ounce/sq yd/day). This is light dependent, with night-time production 90% lower than that during the middle of the day.

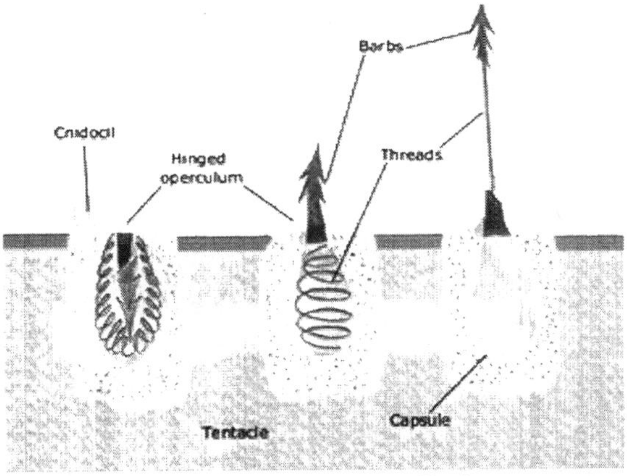

Figure: Nematocyst discharge: A dormant nematocyst discharges response to nearby prey touching the cnidocil, the operculum flap opens, and its stinging apparatus fires the barb into the prey, leaving a hollow filament through which poisons are injected to immobilise the prey, then the tentacles manoeuvre the prey to the mouth.

Tentacles: Nematocysts are stinging cells at the tips of the calices that carry venom, which they rapidly release in response to contact with another organism.

The tentacles also bear a contractile band of epithelium called the pharynx. Jellyfishand sea anemones also carry nematocysts.

Ecology

Feeding: Polyps feed on a variety of small organisms, from microscopic plankton to small fish. The polyp's tentacles immobilise or kill prey using their nematocysts (also known as 'cnidocysts'). The tentacles then contract to bring the prey into the stomach. Once digested, the stomach reopens, allowing the elimination of waste products and the beginning of the next hunting cycle.

Zooxanthellae Symbiote: Many corals, as well as other cnidarian groups such as Aiptasia (a sea anemone) form a symbiotic relationship with a class of algae, zooxanthellae, of the genus *Symbiodinium*. *Aiptasia*, a familiar pest among coral reef aquarium hobbyists, serves as a valuable model organism in the study of cnidarian-algal symbiosis. Typically, each polyp harbours one species of algae. Via photosynthesis, these provide energy for the coral, and aid in calcification. The algae benefit from a safe place to live and consume the polyp's carbon dioxide and nitrogenous waste. Due to the strain the algae can put on the polyp, stress on the coral often drives them to eject the algae. Mass ejections are known as coral bleaching, because the algae contribute to coral's brown colouration; other colours, however, are due to host coral pigments, such as green fluorescent proteins (GFPs). Ejection increases the polyp's chance of surviving short-term stress—they can regain algae, possibly of a different species at a later time. If the stressful conditions persist, the polyp eventually dies.

Reproduction

Corals can be both gonochoristic (unisexual) and hermaphroditic, each of which can reproduce sexually and asexually. Reproduction also allows coral to settle in new areas.

Sexual

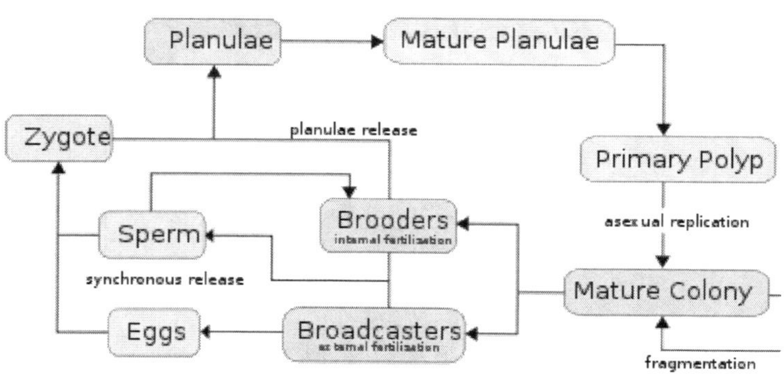

Figure: Life cycles of broadcasters and brooders

Corals predominantly reproduce sexually. About 25% of hermatypic corals (stony corals) form single sex (gonochoristic) colonies, while the rest are hermaphroditic.

Broadcasters: About 75% of all hermatypic corals "broadcast spawn" by releasing gametes—eggs and sperm—into the water to spread offspring. The gametes fuse during fertilization to form a microscopic larva called a planula, typically pink and elliptical in shape. A typical coral colony forms several thousand larvae per year to overcome the odds against formation of a new colony.

Synchronous spawning is very typical on the coral reef, and often, even when multiple species are present, all corals spawn on the same night. This synchrony is essential so male and female gametes can meet. Corals rely on environmental cues, varying from species to species, to determine the proper time to release gametes into the water.

The cues involve temperature change, lunar cycle, day length, and possibly chemical signalling. Synchronous spawning may form hybrids and is perhaps involved in coral speciation. The immediate cue is most often sunset, which cues the release. The spawning event can be visually dramatic, clouding the usually clear water with gametes.

Brooders: Brooding species are most often ahermatypic (not reef-building) in areas of high current or wave action. Brooders release only sperm, which is negatively buoyant, sinking on to the waiting egg carriers who harbour unfertilized eggs for weeks. Synchronous spawning events sometimes occurs even with these species. After fertilization, the corals release planula that are ready to settle.

Planulae: Planulae exhibit positive phototaxis, swimming towards light to reach surface waters, where they drift and grow before descending to seek a hard surface to which they can attach and begin a new colony. They also exhibit positive sonotaxis, moving towards sounds that emanate from the reef and away from open water. High failure rates afflict many stages of this process, and even though millions of gametes are released by each colony, few new colonies form. The time from spawning to settling is usually two to three days, but can be up to two months. The larva grows into a polyp and eventually becomes a coral head by asexual budding and growth.

Asexual

Within a coral head, the genetically identical polyps reproduce asexually, either via gemmation (budding) or by longitudinal or transversal division, both shown in the photo of *Orbicella annularis*.

Budding involves splitting a smaller polyp from an adult. As the new polyp grows, it forms its body parts. The distance between the new and adult polyps grows, and with it, the coenosarc.

Figure: Calices (basal plates) of *Orbicella annularis* showing multiplication by gemmation (small central calice) and division (large double calice)

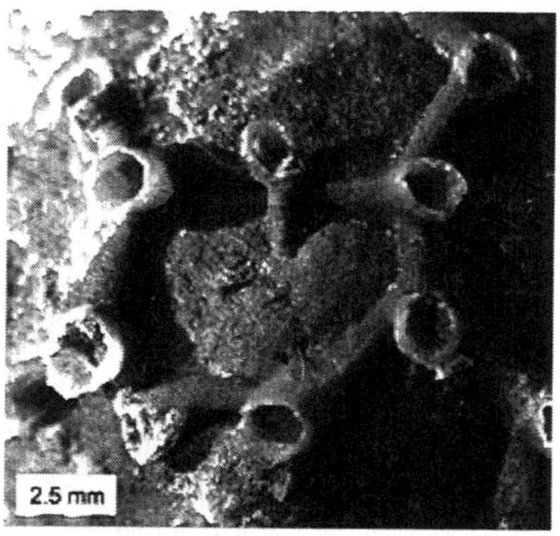

Figure: The tabulate coral *Aulopora* (Devonian) showing initial budding from protocorallite

Budding can be:

- Intratentacular—from its oral discs, producing same-sized polyps within the ring of tentacles
- Extratentacular—from its base, producing a smaller polyp

Division forms two polyps each as large as the original. Longitudinal division begins when a polyp broadens and then divides its coelenteron, analogous to splitting a log along its length. The mouth also divides and new tentacles form. The two "new" polyps then generate their missing body parts and exoskeleton. Transversal division occurs when polyps and the exoskeleton divide transversally into two parts. This means one has the basal disc (bottom) and the other has the oral disc (top), similar to cutting the end off a log. The new polyps must separately generate the missing pieces.

Asexual reproduction has several benefits for these sessile colonial organisms:

- Cloning allows high reproduction rates, supporting rapid habitat exploitation.
- Modular growth allows biomass to increase without a corresponding decrease in surface-to-volume ratio.
- Modular growth delays senescence, by allowing the clone-type to survive the loss of one or more modules.
- New modules can replace dead modules, reducing clone-type mortality and preserving the colony's territory.
- Spreading the clone type to distant locations reduces clone-type mortality from localised threats.

Colony Division: Whole colonies can reproduce asexually, forming two colonies with the same genotype.

- Fission occurs in some corals, especially among the family Fungiidae, where the colony splits into two or more colonies during early developmental stages.
- Bailout occurs when a single polyp abandons the colony and settles on a different substrate to create a new colony.
- Fragmentation involves individuals broken from the colony during storms or other disruptions. The separated individuals can start new colonies.

Reefs

Evolutionary history

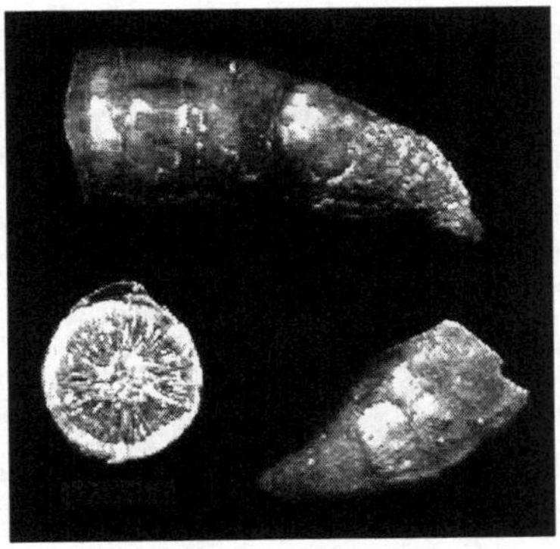

Figure: Solitary rugose coral (*Grewingkia*) in three views; Ordovician, southeastern Indiana

Although corals first appeared in the Cambrian period, some 542 million years ago, fossils are extremely rare until the Ordovician period, 100 million years later, when rugose and tabulate corals became widespread.

Tabulate corals occur in limestones and calcareous shales of the Ordovician and Silurian periods, and often form low cushions or branching masses alongside rugose corals. Their numbers began to decline during the middle of the Silurian period, and they became extinct at the end of the Permian period,250 million years ago. The skeletons of tabulate corals are composed of a form of calcium carbonate known as calcite.

Rugose corals became dominant by the middle of the Silurian period, and became extinct early in the Triassic period. The rugose corals existed in solitary and colonial forms, and were also composed of calcite.

The scleractinian corals filled the niche vacated by the extinct rugose and tabulate species. Their fossils may be found in small numbers in rocks from the Triassic period, and became common in the Jurassic and later periods. Scleractinian skeletons are composed of a form of calcium carbonate known as aragonite. Although they are geologically younger than the tabulate and rugose corals, their aragonitic skeleton is less readily preserved, and their fossil record is less complete.

At certain times in the geological past, corals were very abundant. Like modern corals, these ancestors built reefs, some of which ended as great structures in sedimentary rocks.

Fossils of fellow reef-dwellers algae, sponges, and the remains of many echinoids, brachiopods, bivalves, gastropods, and trilobites appear along with coral fossils. This makes some corals useful index fossils that enabled geologists to date the rocks in which they are found. Coral fossils are not restricted to reef remnants, and many solitary fossils may be found elsewhere, such as *Cyclocyathus*, which occurs in England's Gault clay formation.

A Petoskey stone is a rock and a fossil, often pebble-shaped, that is composed of a fossilized coral, *Hexagonaria percarinata*. They are found predominantly in Michigan's Upper Peninsula, and the northwestern portion of Michigan's Lower Peninsula.

Threats: Coral reefs are under stress around the world. In particular, coral mining, agricultural and urban runoff, pollution (organic and inorganic), overfishing, blast fishing, disease, and the digging of canals and access into islands and bays are localised threats to coral ecosystems. Broader threats are sea temperature rise, sea level rise and pH changes from ocean acidification, all associated with greenhouse gas emissions. In 1998, 16% of the world's reefs died as a result of increased water temperature.

General estimates show approximately 10% of the world's coral reefs are dead. About 60% of the world's reefs are at risk due to human-related activities. The threat to reef health is particularly strong in Southeast Asia, where 80% of reefs are endangered. Over 50% of the world's coral reefs may be destroyed by 2030; as a result, most nations protect them through environmental laws.

In the Caribbean and tropical Pacific, direct contact between ~40 to 70% of common seaweeds and coral causes bleaching and death to the coral via transfer of lipid–solublemetabolites. Seaweed and algae proliferate given adequate nutrients and limited grazing by herbivores such as parrotfish.

Water temperature changes of more than 1-2 degrees Celsius (1.8-3.6 degrees Fahrenheit) or salinity changes can kill coral. Under such environmental stresses, corals expel their zooxanthellae; without them coral tissues reveal the white of their skeletons, an event known as coral bleaching.

Submarine springs found along the coast of Mexico's Yucatán Peninsula produce water with a naturally low pH (a measure of acidity) providing conditions similar to those expected to become widespread as the oceans absorb carbon dioxide. Surveys discovered multiple species of live coral that appeared to tolerate the acidity. The colonies were small and patchily distributed, and had not formed structurally complex reefs such as those that compose the nearby Mesoamerican Barrier Reef System.

Protection: Many governments now prohibit removal of coral from reefs, and inform coastal residents about reef protection and ecology. While local action such as habitat restoration and herbivore protection can reduce local damage, the longer-term threats of acidification, temperature change and sea-level rise remain a challenge. To eliminate destruction of corals in their indigenous regions, projects have been started to grow corals in non-tropical countries.

Relation to Humans

Local economies near major coral reefs benefit from an abundance of fish and other marine creatures as a food source. Reefs also provide recreational scuba diving and snorkeling tourism. These activities can damage coral.

In medicine chemical compounds from corals are used for cancer, AIDS, pain, and other uses. Coral skeletons, e.g., *Isididae* are also used for bone grafting in humans.

Live coral is highly sought after for aquaria. Soft corals are easier to maintain in captivity than hard corals.

Jewellery: Coral's many colours give it appeal for necklaces and other jewellery. Intensely red coral is prized as a gemstone. Sometimes called fire coral, it is not the same as fire coral. Red coral is very rare because of overharvesting.

Construction: Coral reefs on land provide lime for use as building blocks ("coral rag"). Coral *rag* is an important local building material in places such as the East African coast.

Climate Research: The annual growth bands in deep sea bamboo corals (*Isididae*) and others may be among the ocean's first organisms to display the effects of ocean acidification. They produce growth rings similar to those of trees, and can provide a view of changes in the condition in the deep sea over time. They allow geologists to construct year-by-year chronologies, a form of incremental dating, which underlie high-resolution records of past climatic and environmental changes using geochemical techniques. Certain species form communities called microatolls, which are colonies whose top is dead and mostly above the water line, but whose perimetre is mostly submerged and alive. Average tide level limits their height. By analysing the various growth morphologies, microatolls offer a low resolution record of sea level change. Fossilized microatolls can also be dated using radioactive carbon dating. Such methods can help to reconstruct Holocene sea levels.

Aquaculture: Coral aquaculture, also known as *coral farming* or *coral gardening*, is the cultivation of corals for commercial purposes or coral reef restoration. Aquaculture is showing promise as a potentially effective tool for restoring coral reefs, which have been declining around the world. The process bypasses the early growth stages of corals when they are most at risk of dying. Coral seeds are grown in nurseries then replanted on the reef. Coral is farmed by coral farmers who live locally to the reefs and farm for reef conservation or for income. It is also farmed by scientists for research, by businesses for the supply of the live and ornamental coral trade and by private aquarium hobbyists.

Reef Aquarium

Figure: Reef aquarium in Monaco

A reef aquarium or reef tank is a marine aquarium that prominently displays live corals and other marine invertebrates as well as fish that play a role in maintaining the tropical coral reef environment. A reef aquarium requires appropriately intense lighting, turbulent water movement, and more stable water chemistry than fish-only marine aquaria, and careful consideration is given to which reef animals are appropriate and compatible with each other.

Components

Reef aquariums consist of a number of components, in addition to the livestock, including:

Display tank: The primary tank in which the livestock are kept and shown.

Stand: A stand allows for placement of the display tank at eye level and provides space for storage of the accessory components.

Sump: An accessory tank in which mechanical equipment is kept. A remote sump allows for a clutter-free display tank.

Refugium: An accessory tank dedicated to the cultivation of beneficial macroalgae and microflora/fauna. The refugium and sump are often housed in a single tank with a system of dividers to separate the compartments.

Lighting: Several lighting options are available for the reef-keeper and are tailored to the types of coral kept.

Canopy: The canopy houses the light fixtures and provides access to the tank for feeding and maintenance.

Filtration and water movement: A variety of filtration and water movement strategies are employed in reef aquaria. Bulky equipment is often relegated to the sump.

Display Tank

Aquarium: Quascaping, Sump (aquarium), and Refugium (fishkeeping)

fig 1

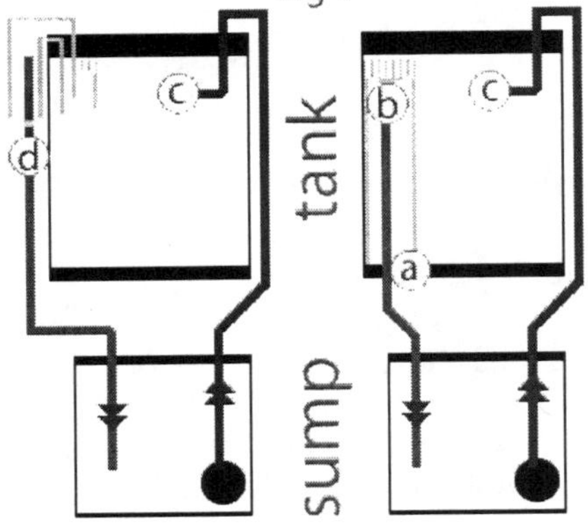

A "reef ready" or simply "drilled" tank is often used. This style of tank has holes drilled into the rear pane allowing water to drain into the sump or refugium. These drains are usually housed in an internal overflow apparatus made of plastic or glass which encloses a drain standpipe and a water return line. The surface water pours over the overflow, down the standpipe, through PVC piping, into the sump. After transiting the sump, water is pushed by a return water pump through the second hole and into the aquarium. Alternatively, standard non-drilled aquariums employ an external "hang-on" overflow that feeds water via continuous siphonto the sump. The tanks are usually constructed from either glass or acrylic. Acrylic has the advantage of optical clarity, lightness, and ease of drilling. Drawbacks include a tendency to scratch easily, bowing, and often limited access from above due to top bracing. Glass aquariums are heavier but harder to scratch. Other materials such as epoxy-coated plywood have been used by industrious DIYers.

Filtration

The primary biological filtration for reef aquariums usually comes from the use of live rock which come from various tropical zones around existing reefs, or more recently aquacultured rock from Florida. Some reef keepers also use what is called deep sand beds (DSB). These are often employed to augment the biological filtration by aiding in the reduction of nitrate, a waste product in an incomplete nitrogen cycle. Deep sand bed opponents may prefer a "bare bottom" or "suspended reef" which allows for easier removal of nitrate-generating accumulated detritus. This biological filtration is usually supplemented by protein skimmers. Protein skimmers use the foam fractionation process wherein air is introduced into a water stream creating microbubbles. Organic waste adheres to the surface of these microbubbles and is removed as it overflows at the reactor surface into a removable cup. This group of elements used in conjunction is characteristic of the Berlin Method, named for the city in which it was first devised.

In recent years the Berlin Method is often supplemented with a refugium. A refugium provides many benefits, which include nitrate reduction, as well as providing a natural food source. It typically houses two main species of macroalgae, including *Caulerpa prolifera* or *chaetomorphae* or both (because these two strains are known to not spore but grow by rooting to propagate). Macroalgae is used for two reasons: to remove from the water excess nutrients such as nitrate, phosphate, and iron, and to support beneficial microflora and fauna (zooplankton). Small invertebrates (copepods and amphipods) are provided a space free of predation to grow and, when returned to the display tank, serve as food for corals and fish. Conventional combined mechanical/biological filtration used in fish only systems is avoided because those filters trap detritus and produce nitrate which may stunt or even kill many delicate corals. Chemical filtration in the form of activated carbon is used when needed to remove discolouration of the water, or to remove dissolved matter (organic or otherwise) to help purify the water in the reef system.

Water Movement

Water movement is important in the reef aquarium with different types of coral requiring different flow rates. At present, many hobbyists advocate a water turnover rate of 10x: 10 x aquarium capacity in gallons = required flow in gallons per hour. This is a general rule with many exceptions. For instance, Mushroom Coral requires little flow and is commonly found in crevices near the base of the reef. Species such as Acropora and Montipora thrive under much more turbulent conditions in the range of 30 to 40 times more flow, which imitates breaking waves in shallow water near the tip of the reef. The directions which water pumps are pointed within an aquarium will have a large effect on flow speeds. Many corals will gradually move themselves to a different area of the tank if the water movement in its current area is not satisfactory.

"Since flow speed is the critical measure for determining the rate of gas exchange, turnover does little to convey how fast a coral will respire and photosynthesize."

Reef ready tanks obtain at least a portion of the required water motion from the pump that returns water from the sump. This flow usually is augmented by other strategies. A popular strategy is placement within the display tank of multiple powerheads. Powerheads are simply small submersible water pumps that produce a laminar or narrow, unidirectional water stream. If the presence of the powerhead in the tank does not fit with the aesthetics of the display, small holes may be drilled in an overflow of a tank and the bulk of the powerhead can be hidden, leaving only the small funnel spout visible in the tank. The pumps may be alternately switched on and off using a wave timer and aimed at one another or at the aquarium glass to create turbulent flow in the tank. Drawbacks to the use of these powerheads include their capacity to clutter the display tank, propensity for excess heat production, and the laminar quality of water flow often produced. Another method is the closed loop in which water is pulled from the main tank into a pump which returns the water back into the aquarium via one or more returns to create water turbulence. Newer submersible propeller pumps are gaining popularity and are able to generate large volumes of turbulent water flow without the intensely directed laminar force of a power head. Propeller pumps are more energy-efficient than powerheads, but require a higher initial investment.

Another recent method is the gyre tank. A gyre tank encourages a maximum amount of water momentum through a divider in the centre of the aquarium. The divider leaves an open, unobstructed space which provides a region with little friction against water movement. Building water momentum using a gyre is an efficient method to increase flow, thus benefiting coral respiration and photosynthesis. Water flow is important to bring food to corals, since no coral fully relies on photosynthesis for food. Gas exchange occurs as water flows over a coral, bringing oxygen and removing gases and shedding material. Water flow assists in reducing the risk of thermal shock and damage by reducing the coral's surface temperature. The surface temperature of a coral living near the water's surface can be significantly higher than the surrounding water due to infrared radiation.

Lighting

With the advent of newer and better technologies, increasing intensities and a growing spectrum, there are many options to consider.

Many, if not most aquarium corals contain within their tissue the symbiotic algae called zooxanthellae. It is these zooxanthellae that require light to perform photosynthesis and in turn produce simple sugars that the corals utilise for food. The challenge for the hobbyist is to provide enough light to allow photosynthesis to maintain a thriving population of zooxanthellae in a coral tissue. Though this may seem simple enough, in reality this can prove to be a very complex task.

Some corals such as mushroom corals and polyp corals require very little light to thrive. Conversely, large-polyp stony corals such as brain coral, bubble coral, elegance coral, cup coral, torch coral, and trumpet coral require moderate amounts of light, and small polyp stony corals such as *Acropora*, *Montipora*, *Porites*, and *Pocillopora* require high intensity lighting.

Of the various types, most popular aquarium lighting comes from metal halide lamps, very high output or VHO, compact fluorescent and T5 high output lighting systems. Although they were once widely used, many reef tank aquarists have abandoned T12 and T8 fluorescent lamps due to their poor intensity, and mercury vapour due to its production of a limited light spectrum.

Recent advances in lighting technology have also made available a completely new technology for aquarium lighting: lightemitting diodes (LEDs). Although LEDs themselves are not new, the technology has only recently been adapted to produce systems with qualities that allow them to be considered viable alternatives to gas- and filament-based aquarium lighting systems. The newness of the technology does cause them to be relatively expensive, but these systems bring several advantages over traditional lighting. Although their initial cost is much higher, they tend to be economical in the long run because they consume less power and have far longer lifespans than other systems. Also, because LED systems are made of hundreds of very small bulbs, a microcomputer can control their output can be controlled to simulate daybreak and sunset. Some systems also have the ability to simulate moonlight and the phases of the moon, as well as vary the colour temperature of the light produced. The choices for aquarium lighting are made complicated by variables such as colour temperature, (measured in kelvins), colour rendering index (CRI), photosynthetically active radiation (PAR) and lumens. Power output available to the hobbyist can range from a meager 9 W fluorescent lamp to a blinding 1000 W metal halide lamp. Lighting systems also vary in the light output produced by each bulb type—listed in order of weakest to strongest they would be: T8/12 or normal output lamps, compact fluorescent and T5 high output, VHO, and metal halide lamps. To further complicate matters, there are several types of ballasts available: electric ballast, magnetic ballast, and pulse start ballast.

Heating and Cooling

Reef tanks are usually kept at a temperature between 25 and 28 °C (75-82 °F). Radical temperature shifts should be avoided as these can be particularly harmful to reef invertebrates and fish. Depending on the location of the tank and the conditions therein (i.e. heat/air conditioning), one may install a heater and/or a chiller for the tank. Heaters are relatively inexpensive and readily available at any local fish store. Aquarists frequently use the sump to hide unsightly equipment such as heaters. Chillers, on the other hand, are expensive and are more difficult to locate. For many aquarists, installing surface fans and running home air conditioning suffice in place of a chiller. Fans cool the tank via evapourative cooling and require more frequent top-off of the aquarium water.

Water Chemistry

Stony corals, which are defined by their calcerous calcium carbonate skeletons ($CaCO_3$), are the focus of many advanced reef keepers. These corals require additional attention to water chemistry, especially maintenance of stable and optimal calcium, carbonate, and pH levels.

These parameters may be tracked and adjusted with test kits and frequent manual dosing of calcium and pH buffer additives requiring no additional equipment. Alternatively, automated methods employing small dedicated computers with electronic water quality monitoring capabilities are often used to control water chemistry parameters via several components including calcium reactors and kalkwasser reactors. Calcium reactors are canisters filled with crushed coral skeletons. Carbon dioxide is injected into the canister acidifying the water and dissolving the coral skeletons. The acidified and $CaCO_3$ rich solution is then pumped into the sump. The excess CO_2 then diffuses out of the water and into the air leaving behind the $CaCO_3$. Kalkwasser is an aqueous solution of calcium hydroxide, $Ca(OH)_2$. The kalk reactor stirs the solution, preventing precipitation, and dispenses the solution into the sump where the $Ca(OH)_2$ combines with dissolved CO_2 to produce $CaCO_3$. These components must be controlled by a computer to prevent dangerous changes in pH due to the acidic calcium reactor effluent or alkaline kalkwasser effluent.

Optimal water parameters are:

- Salinity: 1.022–1.025 sg or 30–34 parts per thousand
- Temperature: 76–80° Fahrenheit
- Ammonia (NH_3): 0 ppm (parts per million)
- Nitrite (NO_2^-): 0 ppm
- Nitrate (NO_3^-): 0–10 ppm
- Phosphate (PO_4^{-3}): 0 ppm
- pH: 8.2–8.6
- Calcium (Ca^{2+}): 400–450 ppm
- Alkalinity: 7–12 dKH

Trace elements can become depleted by marine livestock and filtration, and can be replenished during a water change.

Safety

Large volumes of highly conductive salt water, complex plumbing, and numerous electrical appliances housed in close proximity certainly pose a significant risk of damage to both person and property and require close attention to safety. All equipment should be used according to manufacturer instructions. Electrical equipment should be placed above water level whenever possible, and drip loops should always be used. Circuit limits should never be exceeded and all appliances should be plugged into ground fault circuit interrupter (GFCI) outlets. These can be purchased at any hardware store and are relatively easy to install. Plug in GFCI power strips are also readily available. Home monitoring equipment with water sensors can also be adapted for the home aquarist and used to alert the owner of power outages or water overflows. This equipment can allow for timely intervention in a potential disaster and provides an added sense of security for frequent travellers.

Nano Reefs

A nano reef is a type of marine aquarium that is typically less than 140 litres (30 gallons). The exact limit that distinguishes a nano reef from a regular reef is somewhat ill-defined (some claim that anything less than 180 litres/40 gallons would qualify), but 140 litres (30 gallons) seems to be the generally accepted limit. Nano reefs have become quite popular in recent years among fish keeping hobbyists, primarily because of their smaller size, maintainability, and the possibility of lower costs. The burgeoning interest in this niche of marine aquarium science has fostered several notable contributions ranging from specific consumer products such as specialised aquarium filters, compact high intensity lighting systems and smaller circulation pumps. Such equipment allows the aquarist to maintain an environment wherein many marine organisms are capable of thriving. Nano reefs are very commonly sold as complete kits which contain the tank, stand, power compact T5, T8, PL lamps or metal halide lighting, protein skimmer, UV steriliser, 3 or more stage filtration, a heater and a water pump or powerhead. However, many nano reef keepers decide to upgrade their aquariums with better quality equipment such as a more powerful protein skimmer or lighting.

Pico Reefs

Another term gaining popularity is *pico reef*, which is used to refer to the smallest of nano reef aquariums. Most online forum polls set the range of approximately 5.5 gallons and below as pico reefs. These tiny tanks require even more diligence with regard to water changes and attention to water chemistry because the small water volume provides little room for error.

Care must be exercised when stocking these tiny tanks because too many inhabitants can easily overload the tank's ability to process wastes effectively. For the smallest of pico reefs, even the presence of a single fish is discouraged. Pico reefs often consist of live rock, hardy corals, and small invertebrates such as hermit crabs and marine snails.

The keeping of pico reef aquariums has tested the extent of allelopathy, the chemical and physical means by which corals compete for space. Before the advent of these concentrated environments, it was thought to be impossible for corals of even a few mixed genera to occupy such a small shared water volume.

Challenges Associated with Small Reef Aquariums

Because of the small water volume, nano reef aquariums require extra attention to water quality compared to aquariums of larger water volumes. Many experienced reef aquarists recommend testing the water twice weekly, with water changes at least every week.

In particular, ammonia, nitrite, nitrate, pH, salinity, alkalinity, calcium and phosphate levels should be monitored closely. When it comes to nano reefs, even minute changes in water conditions such as mild temperature fluctuations can be problematic, whereas the greater water volume of larger aquariums provides a more stable and flexible environment.

Nano reefs also require extra care in the selection of occupants. There are two major factors to be considered: biological load, i.e. the ability of the tank to process the wastes produced by the occupants, and species compatibility. These issues, though present in larger tanks, are magnified in the nano tank. Species considered reef safe and able to coexist in larger tanks may not do well in a nano tank due to their close physical proximity. For this reason, smaller species of fish such as gobies and clownfish are popular choices due to their relatively small size and ability to coexist peacefully with other tank inhabitants.

Filtration in Nano Reefs

Many nano reef aquarists prefer their displays to be as natural-looking as possible, and therefore choose to use as few mechanical filtration methods as possible. A primary filtration method in nano reefs is live rock and live sand, which are pieces of rock and sand that have broken from the coral reef and are populated with beneficial bacteria and other organisms that aid in breaking down organic wastes produced by larger organisms in the nano reef. Other nano reef aquarists use devices such as protein skimmers to remove excess waste from the aquarium, before it has a chance to be broken down to nitrate. Removing the excess wastes mechanically can reduce the frequency of water changes needed to keep nitrate levels low. Delaying the action of the mechanical filters, such as by means of a day-night timer, can allow invertebrates to filter-feed naturally. A refugium may also be used to export nutrients, when packed with macroalgae such as *Chaetomorpha*, and live rock. Deep sand bed filters are another filtration method.

Recently there have been several "natural" methods of processing waste in the aquarium and specifically small environments as nano-reefs. The research on the encouragement of the development of different types of sponges and micro-organisms to process the pollutants in the aquarium, a matter that has been gaining popularity in the aquarium community.

Fishkeeping

Fishkeeping is a popular hobby concerned with keeping fish in a home aquarium or garden pond. There is also a fishkeeping industry, as a branch of agriculture.

Types of Fishkeeping Systems

Fishkeepers are often known as "aquarists", since many of them are not solely interested in keeping fish. The hobby can be broadly divided into three specific disciplines according to the type of water the fish tolerate: freshwater, brackish, and marine (also called saltwater) fishkeeping.

Freshwater

Freshwater fishkeeping is by far the most popular branch of the hobby, with even small pet stores often selling a variety of freshwater fish, such asgoldfish, guppies, and angelfish. While most freshwater aquaria are community tanks containing a variety of compatible species, single-species breeding aquaria are also popular. Livebearing fish such as mollies and guppies are among those most easily raised in captivity, but aquarists also regularly breed many types of cichlid, catfish, characin, and killifish.

Many fishkeepers create freshwater aquascapes where the focus is on aquatic plants as well as fish. These aquaria include "Dutch Aquaria", named for European aquarists who designed them. In recent years, one of the most active advocates of the heavily planted aquarium is the Japanese aquarist Takashi Amano.

Garden ponds are in some ways similar to freshwater aquaria, but are usually much larger and exposed to ambient weather. In the tropics, tropical fish can be kept in garden ponds, but in the temperate zone species such as goldfish, koi, and orfe work better.

Saltwater

Marine aquaria are generally more difficult to maintain and the livestock is significantly more expensive. As a result this branch tends to attract more experienced fishkeepers. Marine aquaria can be exceedingly beautiful, due to the attractive colours and shapes of the corals and the coral reef fish they host. Temperate zone marine fish are not as commonly kept in home aquaria, primarily because they do not thrive at room temperature. Coldwater aquaria must provide cooler temperature via a cool room (such as an unheated basement) or a refrigeration device known as a 'chiller'.

Marine aquarists often attempt to recreate a coral reef in their aquaria using large quantities of living rock, porous calcareous rocks encrusted with coralline algae, sponges, worms, and other small marine organisms. Larger corals as well as shrimps, crabs, echinoderms, and mollusks are added later on, once the aquarium has matured, as well as a variety of small fish. Such aquaria are sometimes called reef tanks.

8

Aquarium Species

Brackish Water

Brackish water aquaria combine elements of the other types, with salinity that must stay between that of freshwater and seawater. Brackish water fish come from habitats with varying salinity, such as mangroves and estuaries, and do not thrive if kept permanently in freshwater. Although brackish water aquaria are not necessarily familiar to inexperienced aquarists, many species prefer brackish water, including some mollies, manygobies, some pufferfish, monos, scats, and virtually all the freshwater soles.

The Origins of Fishkeeping

Fish have been raised as food in pools and ponds for thousands of years. Brightly coloured or tame specimens of fish in these pools have sometimes been valued as pets rather than food. Many other cultures kept fish for both functional and decorative purposes.

Ancient Sumerians kept wild-caught fish in ponds, before preparing them for meals. Depictions of the sacred fish of Oxyrhynchus kept in captivity in rectangular temple pools have been found in ancient Egyptian art.

Similarly, Asia has experienced a long history of stocking rice paddies with freshwater fish suitable for eating, including various types of catfish and cyprinid. Selective breeding of carp into today's popular and completely domesticated koi and goldfish began over 2,000 years ago in Japan and China, respectively. The Chinese brought goldfish indoors during the Song Dynasty to enjoy them in large ceramic vessels.

In Medieval Europe, carp pools were a standard feature of estates and monasteries, providing an alternative to meat on feast days when meat could not be eaten for religious reasons.

Marine fish have been similarly valued for centuries. Wealthy Romans kept lampreys and other fish in salt water pools. Tertullian reports that Asinius Celer paid 8000sesterces for a particularly fine mullet. Cicero reports that the advocate Quintus Hortensius wept when a favoured specimen died. Rather cynically, he referred to these ancient fishkeepers as the Piscinarii, the "fish-pond owners" or "fish breeders", for example when saying that *...the rich (I mean your friends the fish-breeders) did not disguise their jealousy of me*.

Aquarium Maintenance

Ideal aquarium ecology reproduces the balance found in nature in the closed system of an aquarium. In practice it is virtually impossible to maintain a perfect balance. As an example, a balanced predator-prey relationship is nearly impossible to maintain in even the largest aquaria. Typically, an aquarium keeper must actively maintain balance in the small ecosystems that aquaria provide.

Balance is facilitated by larger volumes of water which dilute the effects of a systemic shock. For example, the death of the only fish in a 10 litres (2.2 imp gal; 2.6 US gal) tank causes dramatic changes in the system, while the death of that same fish in a 400 litres (88 imp gal; 110 US gal) tank that holds many fish creates only a minor imbalance. For this reason, hobbyists often favour larger tanks when possible, as they require less intensive attention.

A variety of nutrient cycles are important in the aquarium. Dissolved oxygen enters at the surface water-air interface or through the actions of an air pump. Carbon dioxidee scapes into the air. The phosphate cycle is an important, although often overlooked, nutrient cycle. Sulphur, iron, and micronutrients enter the system as food and exit as waste. Appropriate handling of the nitrogen cycle, along with a balanced food supply and consideration of biological loading, is usually enough to keep these nutrient cycles in adequate equilibrium.

Water Conditions

The solute content of water is perhaps the most important aspect of water conditions, as total dissolved solids and other constituents can dramatically impact basic water chemistry, and therefore how organisms interact with their environment. Salt content, or salinity, is the most basic classification of water conditions. An aquarium may have freshwater (salinity below 0.5 PPT), simulating a lake or river environment; brackish water (a salt level of 0.5 to 30 PPT), simulating environments lying between fresh and salt, such as estuaries; and salt water or seawater (a salt level of 30 to 40 PPT), simulating an ocean or sea environment. Even higher salt concentrations are maintained in specialised tanks for raising brine organisms. Several other water characteristics result from dissolved materials in the water, and are important to the proper simulation of natural environments. Saltwater is typically alkaline, while the pH of fresh water varies. "Hardness" measures overall dissolved mineral content; hard or soft water may be preferred. Hard water is usually alkaline, while soft water is usually neutral to acidic. Dissolved organic content and dissolved gases content are also important factors.

Home aquarists typically use modified tap water supplied through their local water supply network. Because of the chlorine used to disinfect drinking water supplies for human consumption, tap water cannot be immediately used. In the past, it was possible to "condition" the water by simply letting the water stand for a day or two, which allows the chlorine to dissipate. However, chloramine became popular in water treatment because it stays longer in the water. Additives are available to remove chlorine or chloramine and suffice to make the water ready. Brackish or saltwater aquaria require the addition of a mixture of salts and other minerals. More sophisticated aquarists may modify the water's alkalinity, hardness, or dissolved content of organics and gases. This can be accomplished by additives such as sodium bicarbonate to raise pH. Some aquarists filter or purify their water using one of two processes: deionisation or reverse osmosis. In contrast, public aquaria with large water needs often locate themselves near a natural water source (such as a river, lake, or ocean) in order to have easy access to water that requires only minimal treatment.

Water temperature forms the basis of one of the two most basic aquarium classifications: tropical vs. cold water. Most fish and plant species tolerate only a limited range of water temperatures: Tropical or warm water aquaria maintain an average temperature of about 25 °C (77 °F) are much more common, and tropical fish are among the most popular aquarium denizens. Cold water aquaria maintain temperatures below the room temperature. More importantly than the range is temperature consistency; most organisms are not accustomed to sudden changes in temperatures, which can cause shock and lead to disease. Water temperature can be regulated with a combined thermometre and heating or cooling unit.

Water movement can also be important in accurately simulating a natural ecosystem. Fish may prefer anything from still water up to swift, simulated currents. Water movement can be controlled through the use of aeration from air pumps, powerheads, and careful design of water flow (such as location of filtration system points of inflow and outflow).

Nitrogen Cycle

The Nitrogen Cycle in an Aquarium: Fish are animals and generate waste as they metabolise food, which aquarists must manage. Fish, invertebrates, fungi, and some bacteria excrete nitrogen in the form of ammonia (which converts to ammonium in acidic water) and must then pass through the nitrogen cycle. Ammonia is also produced through the decomposition of plant and animal matter, including fecal matter and other detritus. Nitrogen waste products become toxic to fish and other aquarium inhabitants above a certain concentration.

The Process

A well-balanced tank contains organisms that metabolise the waste products of other inhabitants. Nitrogen waste is metabolised in aquaria by a type of bacteria known as nitrifiers (genus *Nitrosomonas*). Nitrifying bacteria metabolise ammonia into nitrite. Nitrite is also highly toxic to fish in high concentrations. Another type of bacteria, genus *Nitrospira*, on–converts nitrite into less–toxic nitrate. (*Nitrobacter* bacteria were previously believed to fill this role, and appear in "jump start" kits. While biologically they could theoretically fill the same niche as Nitrospira, it has recently been found that *Nitrobacter* are not present in detectable levels in established aquaria, while *Nitrospira* are plentiful.) This process is known in the aquarium hobby as the nitrogen cycle.

Aquatic plants also metabolise ammonia and nitrate, using it to build biomass. However, this is only temporary, as the plants release nitrogen back into the water when older leaves die off and decompose.

Maintaining the Nitrogen Cycle

Although called the nitrogen "cycle" by hobbyists, in aquaria the cycle is not complete: nitrogen must be added (usually indirectly through food) and nitrates must be removed at the end. Nitrogen bound up in plant matter is removed when the plant grows too large.

Hobbyist aquaria typically do not have the requisite bacteria needed to detoxify nitrogen waste. This problem is most often addressed through filtration. Activated carbon filters absorb nitrogen compounds and other toxins from the water.

Biological filters provide a medium specially designed for colonisation by the desired nitrifying bacteria. Activated carbon and other substances, such as ammonia absorbing resins, stop working when their pores fill, so these components have to be replaced with fresh stocks periodically.

New aquaria often have problems associated with the nitrogen cycle due to insufficient beneficial bacteria, which is known as "New Tank Syndrome". Therefore, new tanks have to mature before stocking them with fish. There are three basic approaches to this: the *fishless cycle*, the *silent cycle*, and *slow growth*.

- Tanks undergoing a "fishless cycle" have no fish. Instead, the keeper adds ammonia to feed the bacteria. During this process, ammonia, nitrite, and nitrate levels measure progress.

- The "silent cycle" involves adding fast-growing plants and relying on them to consume the nitrogen, filling in for the bacteria work until their number increases. Anecdotal reports indicate that such plants can consume nitrogenous waste so efficiently that the ammonia and nitrite spikes that occur in more traditional cycling methods are greatly reduced or indetectable.
- "Slow growth" entails slowly increasing the fish population over 6 to 8 weeks, giving bacteria time to grow and reach a balance with the increasing waste production.

Adding too many fish too quickly or failing to allow enough time for the bacteria colony to establish itself in the filter media can lead to ammonia stress. This is not always fatal but can result in the death of aquarium fish. A few days after adding hardy fish for the cycling process, it is essential to look out for the key signs of ammonia stress. These include a lack of movement and appetite, inflammation and redness of the gills, fins, and body, and occasionally gasping for air at the water's surface. The latter can also be attributed to poor aeration, which can be negated by inclusion of an air pump or spray bar in the setup. The largest bacterial populations inhabit the filter; efficient filtration is vital. Sometimes, simply cleaning the filter is enough to seriously disturb the aquarium's balance. Best practice is to flush mechanical filters using compatible water to dislodge organic materials, while preserving bacteria populations. Another safe practice involves cleaning only one half of the filter media every time the filter or filters are serviced to allow the remaining bacteria to repopulate the cleaned half.

Tank Capacity

Biological loading is a measure of the burden placed on the aquarium ecosystem by its living inhabitants. Higher biological loading represents a more complicated ecology, which makes equilibrium easier to perturb. The surface area of water exposed to air limits dissolved oxygen. The population of nitrifying bacteria is limited by the available physical space.

Tank Size

Fish capacity is a function of aquarium size. Limiting factors include the availability of oxygen in the water and the rate at which the filter can process waste. Aquarists apply rules of thumb estimating appropriate population size; the examples below are for small freshwater fish. Larger freshwater fish and most marine fishes need much more generous allowances. Some aquarists claim that increasing water depth beyond some relatively shallow minimum does not affect capacity.

- 1.5 litres of water for each centimetre of fish length (1 gallon per inch).
- 30 square centimetres of surface area per centimetre of fish length (12 square inches per inch).

Experienced aquarists warn against mechanically applying these rules because they do not consider other important issues such as growth rate, activity level, social behaviour, and so on. Once the tank nears capacity, best practice is to add the remaining fish over a period of time while monitoring water quality.

The capacity can be improved by surface movement and water circulation such as through aeration, which not only improves oxygen exchange, but also the decomposition of waste materials.

Other Factors

Other variables affect tank capacity. Smaller fish consume more oxygen per unit of body weight than larger fish. Labyrinth fish can breathe atmospheric oxygen and need less surface area (however, some are territorial, and do not tolerate crowding). Barbs require more surface area than tetras of comparable size. The presence of waste materials presents itself as a variable as well. Decomposition consumes oxygen, reducing the amount available for fish. Oxygen dissolves less readily in warmer water, while warmer water temperature increase fish activity levels, which in turn consume more oxygen.

Fishkeeping Industry

Worldwide, the fishkeeping hobby is a multi-billion dollar industry. The United States is the largest market, followed by Europe and Japan. In 1993 the United States Census Bureau found that 10.6% of U.S. households owned ornamental freshwater or saltwater fish, with an average of 8.8 fish per household. In 1993, the retail value of the fish hobby in the United States was US$910 million. In 2002, census data indicated that aquarium products and fished accounted for USD684 million dolllars.

From 1989 to 1992, almost 79% of all U.S. ornamental fish imports came from Southeast Asia and Japan. Singapore, Thailand, the Philippines, Hong Kong, and Indonesia were the top five exporting nations. South America was the second largest exporting region, accounting for 14% of the total annual value. Colombia, Brazil and Peru were the major suppliers.

Approximately 201 million fish worth $44.7 million were imported into the United States in 1992. These fish comprised 1,539 different species; 730 freshwater species, and 809 saltwater species. Freshwater fish accounted for approximately 96% of the total volume and 80% of the total import value. Only 32 species had import values over $10,000. The top species were freshwater and accounted for 58% of the total imported value. The top imported species are the guppy, neon tetra, platy, betta, Chinese algae eater, and goldfish. Given 91.9 million total US households in 1990, 9.7 million are fishkeepers. 8.8 fish per household implies a total aquarium fish population of approximately 85.7 million, suggesting that the US aquarium fish population turns over more than 2.3 times per year, counting only imported fish. Historically, fish and plants for the first modern aquaria were gathered from the wild and transported (usually by ship) to Europea and America. During the early 20th century many species of small colourful tropical fishwere exported from Manaus, Brazil; Bangkok, Thailand; Jakarta, Indonesia; the Netherlands Antilles; Kolkata, India; and other tropical countries. Import of wild fish, plants, and invertebrates for aquaria continues today around the world. Many species have not been successfully bred in captivity. In many developing countries, locals survive by collecting specimens for the aquarium trade., and continue to introduce new species to the market.

Fish Breeding

Fish breeding is a challenge that attracts many aquarists. While some species reproduce freely in community tanks, most require special conditions, known as spawning triggers before they will breed. The majority of fish lay eggs, known as spawning, and the juvenile fish that emerge are very small and need tiny live food or substitutes to survive. A fair number of popular aquarium fish are livebearers which produce a small number of relatively large offspring. These usually take ground flake food straight away.

Animal Welfare

A properly maintained aquarium allows fish to socialise with their own species and in many cases breed successfully. This is in marked contrast to the conditions of larger animals like cats and dogs, which are often kept alone and neutered in an environment different from what they would experience in the wild. However, fish are often maintained in inadequate conditions. Therefore they live short lives and never breed.

Inexperienced aquarists often keep too many fish in one tank, or add fish too quickly into an immature aquarium, killing many of them. This has given the hobby a bad reputation among some animal welfare groups, such as PETA, who accuse aquarists of treating aquarium fish as cheap toys to be replaced when they die.

Goldfish and bettas in particular have often been kept in small bowls or aquaria that are too small for their needs. In some cases fish have been installed in all sorts of inappropriate objects such as the "AquaBabies Micro Aquaria", "Bubble Gear Bubble Bag", and "Betta in a Vase", all of which house live fish in unfiltered and insufficient water. The latter is sometimes marketed as a complete ecosystem because a plant is included in the neck of the vase. Some sellers claim the fish eat the plant roots. However, bettas are carnivorous and need live food or pellet foods. They cannot survive on plant roots. Another problem is that the plant sometimes blocks the betta's passage to the water surface. They are labyrinth fish, and need to breathe at the surface to avoid suffocation.

Such products are aimed at people looking for a novelty gift. Aquarists actively condemn them. Similarly, the awarding of goldfish as prizes at funfairs is traditional in many parts of the world, but has been criticized by aquarists and activists as cruel and irresponsible. The United Kingdom outlawed live-animal prizes such as goldfish in 2004. The use of live prey to feed carnivorous fish such as piranhas also draws criticism.

Fish Modification

Modifying fish to make them more attractive as pets is increasingly controversial. Historically, artificially dyeing fish was common. Glassfish in particular were often injected with fluorescent dyes. The major British fishkeeping magazine, Practical Fishkeeping, has been effective in its campaign to remove these fish from the market by educating retailers and aquarists to the cruelty and health risks involved.

In 2006, *Practical Fishkeeping* published an article exposing the techniques for performing cosmetic surgery on aquarium fish, without anaesthesia, as described by Singaporean fishkeeping magazine *Fish Love Magazine*. The tail is cut off and dye is injected into the body. The piece also included the first documented evidence to demonstrate that parrot cichlids are injected with coloured dye. Hong Kong suppliers were offering a service in which fish could be tattooed with company logos or messages using a dye laser; such fishes have been sold in the UK under the name of Kaleidoscope gourami and Striped parrot cichlid. Some people give their fish body piercings.

Hybrid fish such as flowerhorn cichlids and parrot cichlids are highly controversial. Parrot cichlids in particular have a very unnatural shape that prevents them from swimming properly and makes it difficult for them to engage in normal feeding and social behaviours. The biggest concern with hybrids is that they may be bred with native species, making it difficult for hobbyists to identify and breed particular species. This is especially important to hobbyists who shelter species that are rare or extinct in the wild. Extreme mutations have been selected for by some breeders; some fancy goldfish varieties in particular have features that prevent the fish from swimming, seeing, or feeding properly.

Genetically modified fish such as the GloFish are likely to become increasingly available, particularly in the United States and Asia. Although GloFish are said to be unharmed by their genetic modifications, they remain illegal in many places, including the European Union, though at least some have been smuggled into the EU, most likely from Taiwan, via the Czech Republic.

Conservation

The two main sources of fish for aquaria are from capture in the wild or captive breeding. United Nations studies show that more than 90% of freshwater aquarium fish are captive–bred, while virtually all marine fish and invertebrates are caught. The few marine species bred in captivity supplement but rarely replace the trade in wild-caught specimens. Wild-caught animals provide valuable income for people in regions lacking other high-value exports.

Marine fish are typically less resilient during transport than freshwater fish, and relatively large numbers of them die before they reach the aquarist. Although the aquarium trade probably represents a minor threat to coral reefs compared to habitat destruction, fishing for food, and climate change, it is a booming trade and may be a serious problem in specific locations such as the Philippines and Indonesia where most collecting is done. Catching fish in the wild can reduce their population sizes, placing them in danger of extinction in collecting areas, as has been observed with the dragonet *Synchiropus splendidus*.

Collecting

In theory, reef fish should be a good example of a renewable resource that encourages fishermen to maintain the integrity and diversity of the natural habitat: more and better fish can be exported from pristine habitats than those that have been polluted or over-harvested. However, this has not been the case in similar industries such as fur trapping, logging, or fishing that experience the *tragedy of the commons*. Fish are caught by net, trap, or cyanide. Collecting expions can be lengthy and costly, and are not always successful. Fish can also be injured during collection and/or shipping; mortality rates during shipping are high. Many others are weakened by stress and become diseased.

Other problems include the poisoning of coral reefs and non-target species, the depletion of rare species from their natural habitat, and ecosystem degradation from large scale removal of key species. Additionally, destructive fishing techniques concern environmentalists and hobbyists alike. There has been a concerted movement to captive breeding and certification programs for wild-caught fish. Two thirds of American marine aquarists surveyed in 1997 preferred farmed coral, and over 80% think that only sustainably caught or farmed fish should be traded.

Cyanide

Cyanide is a poison that stuns and immobilises fish. Fishers put cyanide in the ocean, to ease the process of netting them. It can irreversibly damage or kill the target fish, as well as other fish, mammals, reptiles or invertebrates that are left behind. Some UK-based wholesalers advertise that they avoid cyanide-caught animals. In the Philippines, overfishing and cyanide caused a drastic decline in aquarium fish, leading the country away from cyanide and towards netting healthy animals. Cyanide is also often used for collecting freshwater species, especially in muddy water with lots of vegetation, which complicates catching small and fast moving fish.

Captive Breeding and Aquaculture

Since the Siamese fighting fish (*Betta splendens*) was first successfully bred in France in 1893, captive spawning and brooding techniques used in aquaculture have slowly improved. Captive breeding for aquaria is concentrated in southern Florida, Singapore, Hong Kong and Bangkok, with smaller industries in Hawaii and Sri Lanka. Captive breeding of marine organisms has been in development since the mid-1990s. Breeding for freshwater species is more advanced than for saltwater species. Currently, only a few captive-bred marine species are in the trade, including clownfish, damselfish, and dwarf angelfish.

Aquaculture can help in lessening the impacts on wild stocks, either by using cultivated organisms for sale or by releasing them to replenish wild stocks. Breeding programs help preserve species that have become rare or extinct in the wild, most notably the Lake Victoria cichlids.

Some species have also become important as laboratory animals. Cichlids and poecilids are especially important for studies on learning, mating, and social behaviour. Hobbyists also keep and observe many fishes not otherwise studied, and thereby provide valuable ecological and behavioural data.

Captive breeding has reduced prices for hobbyists, but cultured animals remain more expensive. Selective breeding has also led to wider intra–species variation, creating more diverse commercial stocks.

Invasive Species

Serious problems can occur when fish originally raised in ponds or aquaria are released into the wild. While tropical fish do not survive in temperate climates, they can thrive in waters that are similar to their native habitat. Non–native species that become established are called exotic species. Freshwater examples include various cichlids in Florida, goldfish in temperate waters, and South American suckermouth catfishes in warm waters around the world. Invasive species can seriously disrupt their new homes by preying on, or competing with, native species. Many marine fish have also been introduced into non-native waters, disrupting the local habitat.

Humane Treatment

In January 2011 the Maui County Council just passed measure requiring the aquarium fishery to adopt humane practices for preparing fish for transport to market. The regulations control harvesting and shipping practices, including prohibiting clipping the fins on fish to protect the plastic shipping bags, outlawing puncturing swim bladders that fish use to regulate their buoyancy, which enabled divers to rapidly surface and prohibiting "starving" the fish which permitted smaller shipping bags without killing the fish with their own waste. The measure also requires that shippers file mortality reports on the animals they ship.

Columnaris

Columnaris is a symptom of disease in fish which results from an infection caused by the gram negative, aerobic, rod shaped bacterium *Flavobacterium columnare*. It was previously known by the names of bacillus columnaris, chondrococcus columnaris, cytophaga columnaris and flexibacter columnaris. The bacteria is ubiquitous in fresh water, and cultured fish reared in ponds or raceways are the primary concern – with disease most prevalent in air temperatures above 12–14p C. The disease is highly contagious and the outcome is often fatal. It is not zoonotic.

Figure: Columnaris in a Chinook salmon

Causes

The bacterium usually enters fish through gills, mouth, or small wounds, and is prevalent where high bio-load exists, or where conditions may be stressful due to overcrowding or low dissolved oxygen levels in the water column. The bacteria can persist in water for up to 32 days when the hardness is 50 ppm or more.

Symptoms

An infection will usually first manifest in fish by causing frayed and ragged fins. This is followed by the appearance of ulcerations on the skin, and subsequent epidermal loss, identifiable as white or cloudy fungus-like patches – particularly on the gill filaments. Mucus often also accumulates on the gills, head and dorsal regions. Gills will change colour, either becoming light or dark brown, and may also manifest necrosis. Fish will breathe rapidly and laboriously as a sign of gill damage. Inappetance and lethargy are common, as are mortalities – especially in young fish.

Diagnosis

Bacteria can be isolated from gills, skin and the kidneys. For definitive diagnosis the pathogen should then be cultured on reduced nutrient agar. Inhibiting contaminant growth on the agar by adding antibiotics and keeping the temperature at 37°C should improve culture results. Colonies are small, 3–4 mm in diametre and grow within 24 hours. They are characteristically rhizoid in structure and pale yellow in colour.

Prognosis

Ulcerations develop within 24 to 48 hours. Fatality occurs between 48 to 72 hours if no treatment is pursued; however, at higher temperatures death may occur within hours. Other symptoms may accompany the disease including lethargy, colour loss, redness around the infection site, loss of appetite and twitching or rubbing the body against objects.

Treatment

As Flavobacterium columnare is a gram-negative bacteria, fish can be treated with a combination of the antibiotics Furan-2 and Kanamycin administered together. Medicated food containing oxytetracycline is also an effective treatment for internal infections, but resistance is emerging.

Potassium permanganate, copper sulphate and hydrogen peroxide can also be applied externally to adult fish and fry but can be toxic at high concentrations. Vaccines can also be given in the face of an outbreak or to prevent disease occurrence.

Cryptocaryon

Cryptocaryon irritans (also known as marine white spot disease or marine ich) is a species of ciliate protozoa that parasitizes marine fish, and is one of the most common causes of disease in marine aquaria.

Taxonomy

Cryptocaryon irritans was originally classified as Ichthyophthirius marinus, but it is not closely related to the other species. It belongs to the class Prostomatea, but beyond that its placement is still uncertain.

Clinical

The symptoms and life-cycle are generally similar to those of *Ichthyophthirius* in freshwater fish, including white spots, on account of which *Cryptocaryon* is usually called marine ich. However, *Cryptocaryon* can spend a much longer time encysted. Fish that are infected with Cryptocaryon may have small white spots, nodules, or patches on their skin, fins, or gills. They may also have ragged fins, cloudy eyes, pale gills, increased mucus production, or changes in skin colour, and they may appear thin. Behaviourally, fish may act different. They may scratch, swim abnormally, act lethargic, hang at the surface or on the bottom, or breathe more rapidly as if they are in distress.

Useful treatments of *Cryptocaryon irritans* are copper solutions, formalin solutions and quinine based drugs (such as Chloroquine Phosphate and Quinine Diphosphate).

Infections can be extremely difficult to treat because of other creatures, such as corals and other invertebrates, which will not survive standard treatments. Ideally fish with *Cryptocaryon* are quarantined in a hospital tank, where they can be treated with a copper salt or using hyposalinity. The display tank needs to be kept clear of fish for 6-9 weeks, the longer the better. This gives time for the encysted tomonts to release infectious theronts, which die within 24-48 hours when they cannot find a host.

Dactylogyrus

Dactylogyrus is a genus of the Dactylogyridae family. They are commonly known as gill flukes

Like other monogeneans, *Dactylogyrus* only has one host required to complete its life cycle.

Introduction

Dactylogyrus (common name: Gill Fluke) are oviparous monogeneans that have two pairs of anchors. These anchors can be used to latch onto the gills of a host, particularly freshwater fish such as carp. In heavily infected fish, *Dactylogyrus* can also be found on the buccal cavity. Other characteristics of the *Dactylogyrus* include the appearance of four eye-spots, 14 marginal hooks, one to two connective bars and two needle-like structures and spindle-shaped dactylogyrid-type seminal vesicles.

Life Cycle

The *Dactylogyrus* life cycle is direct, having no intermediate host. The hermaphroditic adults are oviparous and produce eggs into the water which hatch prior to attaching to the gills of a fish host and developing into an onchomiricidium.

After the eggs hatch, water currents aid the free-swimming ciliated larva in reaching its host. The time required for egg maturation into the adult form is temperature dependent. Water temperatures of 72–75°F allow life cycle completion in a few days, whereas temperatures of 34–36°F extend the generation time to five or six months.

Prevalence

Dactylogyrus is a monogean parasite that is usually found on the gills of *Cyprinidae* fishes. The prevalence of *Dactylogyrus* infection on fish differ depending on the seasons. It was found that *Dactylogyrus* infections are at their greatest during late autumn or early winter. Correlation has also been found between the temperature of the water and the intensity of *Dactylogyrus* infection. It is also generally accepted that fish are exposed to increased *Dactylogyrus* infections during their spawning period.

Symptoms

Cyprinidae that are infected by *Dactylogyrus* may have symptoms that include inflamed gills, excessive mucous secretions and accelerated respiration. The infected fish also becomes lethargic, swims near the surface, and its appetite decreases. Additionally the infected fish may hold its gill covers open and scratch its gills on rocks.

In severe infections, *Dactylogyrus* can cause hemorrhaging and metaplasia of the gills which can lead to secondary bacterial infections and death. Heavily infected fish are also anorexic and can be found gasping for air and exhibiting abnormal behaviour such as jumping out of the water.

Treatment

A primary method for control of *Dactylogyrus* is the application of chemicals. Treatment include salt baths, formalin or organophosphates, Bromex-50 and potassium permanganate.

Fin Rot

Fin rot is a symptom of disease or the actual disease in fish. This is a disease which is most often observed in aquaria and aquaculture, but can also occur in natural populations.

Fin rot can be the result of a bacterial infection (*Pseudomonas fluorescens*, which causes a ragged rotting of the fin), or as a fungal infection (which rots the fin more evenly and is more likely to produce a white 'edge'). Sometimes, both types of infection are seen together. Infection is commonly brought on by bad water conditions, injury, poor diet, or as a secondary infection in a fish which is already stressed by other disease.

Fin rot starts at the edge of the fins, and destroys more and more tissue until it reaches the fin base. If it does reach the fin base, the fish will never be able to regenerate the lost tissue. At this point, the disease may attack the fish's body directly.

Fin rot is very common in Bettas due to poor water conditions in pet stores.

Symptoms:

- Fin edges turn black / brown
- Fins fray
- Base of fins inflamed
- Entire fin may rot away or fall off in large chunks

Ways to treat:

- Change the water
- Treat with herbal remedies like Melafix
- Use common salt
- Find out the pH and correct it

Use medication specifically for fin rot:

- Maracyn TC (for positive bacterial infections)
- Maracyn 2 (for negative bacterial infections)
- Melafix (as a preventative) *Melafix contains tea tree oil which in fact damages the labrynth organ in air-breathing fish like bettas. Melafix and bettafix are not recommended for use on bettas.

Prevention:

- Make sure the water is good quality
- Feed fresh food in small portions
- Maintain constant water temperature

Trematoda

Trematoda is a class within the phylum Platyhelminthes that contains two groups of parasitic flatworms, commonly referred to as "flukes".

Taxonomy and Biodiversity

The trematodes or flukes are estimated to include 18,000 to 24,000 species, and are divided into two subclasses. Nearly all trematodes are parasites of mollusks and vertebrates.

The smaller Aspidogastrea, comprising about 100 species, are obligate parasites of mollusks and may also infect turtles and fish, including cartilaginous fish.

The Digenea, which constitute the majority of trematode diversity, are obligate parasites of both mollusks and vertebrates, but rarely occur in cartilaginous fish.

Formerly the Monogenea were included in Trematoda on the basis that these worms are also vermiform parasites, but modern phylogenetic studies have raised this group to the status of a sister class within the Platyhelminthes, along with the Cestoda.

Anatomy

Trematodes are flattened oval or worm-like animals, usually no more than a few centimetres in length, although species as small as 1 millimetre (0.039 in) and as large as 7 centimetres (0.23 ft) are known. Their most distinctive external feature is the presence of two suckers, one close to the mouth, and the other on the underside of the animal. The body surface of trematodes comprises a tough syncitial tegument, which helps protect against digestive enzymes in those species that inhabit the gut of larger animals. It is also the surface of gas exchange; there are no respiratory organs.

The mouth is located at the forward end of the animal, and opens into a muscular, pumping pharynx. The pharynx connects, via a short oesophagus, to one or two blind-ending caeca, which occupy most of the length of the body. In some species, the caeca are themselves branched. As in other flatworms, there is no anus, and waste material must be egested through the mouth.

Although the excretion of nitrogenous waste occurs mostly through the tegument, trematodes do possess an excretory system, which is instead mainly concerned with osmoregulation. This consists of two or more protonephridia, with those on each side of the body opening into a collecting duct. The two collecting ducts typically meet up at a single bladder, opening to the exterior through one or two pores near the posterior end of the animal.

The brain consists of a pair of ganglia in the head region, from which two or three pairs of nerve cords run down the length of the body. The nerve cords running along the ventral surface are always the largest, while the dorsal cords are present only in the Aspidogastrea. Trematodes generally lack any specialised sense organs, although some ectoparasitic species do possess one or two pairs of simple ocelli.

Reproductive System

Most trematodes are simultaneous hermaphrodites, having both male and female organs. There are usually two testes, with sperm ducts that join together on the underside of the front half of the animal. This final part of the male system varies considerably in structure between species, but may include sperm storage sacs and accessory glands, in addition to the copulatory organ, which is either eversible, and termed a *cirrus*, or non-eversible, and termed a penis.

There is usually only a single ovary, which is connected, via a pair of ducts to a number of *vitelline glands* on either side of the body, that produce yolk cells. Eggs pass from the ovary into a glandular receptacle called the *ootype* or *Mehlis' gland*, where fertilization occurs. This opens into an elongated uterus that opens to the exterior close to the male opening. The ovary is often also associated with a storage sac for sperm, and a copulatory duct termed *Laurer's canal*.

Life Cycles

Almost all trematodes infect mollusks as the first host in the life cycle, and most have a complex life cycle involving other hosts. Most trematodes are monoecious and alternately reproduce sexually and asexually. The two main exceptions to this are the Aspidogastrea, which have no asexual reproduction, and the schistosomes, which are dioecious.

In the definitive host, in which sexual reproduction occurs, eggs are commonly shed along with host feces. Eggs shed in water release free-swimming larval forms that are infective to the intermediate host, in whicha sexual reproduction occurs.

A species that exemplifies the remarkable life history of the trematodes is the bird **fluke**, *Leucochloridium paradoxum*. The definitive hosts, in which the parasite multiplies, are various woodland birds, while the hosts in which the parasite grows (intermediate host) are various species of snail. The adult parasite in the bird's gut produces eggs and these eventually end up on the ground in the bird's faeces. Some very fortunate eggs get swallowed by a snail and here they hatch into tiny, transparent larva (miracidium). These larvae grow and take on a sac-like appearance. This stage is known as the sporocyst and it forms a central body in the snail's digestive gland that extends into a brood sac in the snail's head, muscular foot and eye-stalks. It is in the central body of the sporocyst where the parasite replicates itself, producing lots of tiny embryos (redia). The seembryos move to the brood sac and mature into cercaria.

Infections

Human infections are most common in Asia, Africa, South America, or the Middle East. However, trematodes can be found anywhere that human waste is used as fertilizer.

Etymology

Trematodes are commonly referred to as *flukes*. This term can be traced back to the Old English name for flounder, and refers to the flattened, rhomboidal shape of the worms.

The flukes can be classified into two groups, on the basis of the system which they infect in the vertebrate host.

- Tissue flukes infect the bile ducts, lungs, or other biological tissues. This group includes the lung fluke, *Paragonimus westermani*, and the liver flukes, *Clonorchis sinensis* and *Fasciola hepatica*.
- Blood flukes inhabit the blood in some stages of their life cycle. Blood flukes include species of the genus *Schistosoma*.

They may also be classified according to the environment in which they are found. For instance, pond flukes infect fish in ponds.

Ichthyophthirius Multifiliis

Ichthyophthirius multifiliis (commonly known as freshwater white spot disease, freshwater ich, or freshwater ick) is a common disease of freshwater fish. It is caused by the protozoa *Ichtyopthirius*. Ich is one of the most common and persistent diseases. The protozoan is an ectoparasite. White nodules that look like white grains of salt or sugar of up to 1 mm appear on the body, fins and gills. Each white spot is an encysted parasite. It is easily introduced into a fish pond tank, or home aquarium by new fish or equipment which has been moved from one fish-holding unit to another. Once the organism gets into a large fish culture facility, it is difficult to control due to its fast reproductive cycle and its unique life stages. If not controlled, there is a 100% mortality rate of fish. With careful treatment, the disease can be controlled but the cost is high in terms of lost fish, labour, and cost of chemicals.

Whitespot is very damaging to the gills and skin. In heavily infected fish it can cause a rapid loss of condition, considerable distress and death. Infected fish have small white spots on the skin and gills and produce excess mucus, due to irritation. Whitespot causes most damage when entering and leaving the tissues of the fish. This can lead to the loss of skin and ulcers. These wounds can harm the ability of a fish to control the movement of water into its body. Damage caused to the gill tissue of an infected fish can also reduce respiratory efficiency. This means it is more difficult for the fish to obtain oxygen from the water, and becomes less tolerant to low levels of dissolved oxygen.

Description

I. multifiliis is one of the most prevalent protozoan parasites of fish and is an obligate parasite which means that the parasite cannot survive unless live fish are present. Adult organisms have an oval to round shape and measure 0.5 to 1.0 mm in size. The adult parasite is uniformly ciliated and contains a horseshoe-shaped nucleus which can be seen in older individuals. It is an important pathogen of ornamental and farm-raised food fish species when reared under intensive conditions. Wild fish populations are also susceptible and outbreaks are occasionally seen. There are few aquarists that have not met it on one or more occasions.

The ich protozoa goes though the following life stages:

- *Feeding stage:* The ich trophozoite (a protozoan in active stage of life) feeds in a nodule formed in the skin or gill epithelium.
- After it feeds within the skin or gills, the trophozoite falls off and enters an encapsulated dividing stage (tomont). The tomont adheres to plants, nets, gravel or other ornamental objects in the aquarium.
- The tomont divides up to 10 times by binary fission, producing infective theronts, thus dividing rapidly and attacking the fish.

This life cycle is highly dependent on water temperature, and the entire life cycle takes from approximately 7 days at 25 °C (77 °F) to 8 weeks at 6 °C (43 °F).

Marine ich is a similar disease caused by a different ciliate, *Cryptocaryon irritans*.

Life Cycle

The white spots that can be seen on the infected hosts are trophonts. This is the feeding and growing stage. Once it has reached maturity the trophont leaves the host. Now it is known as a tomont. The tomont becomes encysted, producing a sticky capsule. This enables it to attach to any substrate that it comes into contact with, from weeds and stones to fishing equipment, such as line and nets. Within its cyst the tomont divides many times, producing up to 3000 tomites. The tomites break out of the cyst wall and are now theronts. The theronts are heavily ciliated and actively seek out a host, without which they can survive for up to 48 hours. On finding a host the theront penetrates through the skin and develops into a trophont.

Predisposing Factors

There is no dormant stage in the lifecycle. Ich does not lie in wait for a weakened fish to infect. However, any factor that reduces immunity like changes in water temperature and quality may, in a subclinically infected fish, accelerate an outbreak of Ich. The presence of ammonia, nitrite and high levels of nitrate in water does not in itself cause clinical cases of Ich. However, poor water quality will stress fish, allow an outbreak to spread rapidly and increase mortality rates in infected fish.

It has also been shown that other abiotic factors can increase both fish and tadpole susceptibility to ich. These factors include, decreased temperature, predatory cues and increased levels of UV-B radiation.

Diagnosis

The diagnosis of "Ich" can easily be confirmed by microscopic examination of skin and gills. Remove several "white spots" from an infected fish, then mount them on a microscope slide with a few drops of water and cover with glass. The mature parasite is large, dark in colour (due to the thick cilia covering the entire cell), and has a horseshoe-shaped nucleus which is sometimes visible under 100x magnification. The adult parasite moves slowly in a tumbling manner and is easily recognised. The immature forms tomites are smaller, translucent, and move quickly.

Typical behaviours of clinically infected fish include:

- Anorexia (loss of appetite, refusing all food, with consequential wasting)
- Rapid breathing
- Hiding abnormally/ not schooling
- Resting on the bottom

- Flashing
- Rubbing and scratching against objects

A subclinically infected fish will not show any of these signs. For example, a healthy fish with a newly attached trophozoite will not yet have clinical disease. The trophozoite will not become visible to the naked eye until it has fed on the fish and grown to one or two millimetres. A trophozoite attached to the gills usually is not readily seen. A subclinically infected fish may initially only have a single trophozoite.

Skin

Visible Ich lesions are usually seen as one or several characteristic white spots on the body or fins of the fish. The white spots are single cells called trophozoites or trophonts, which feed on the tissues of the host and may grow to 1 mm in diametre. A smear should show ciliates if white spot is present.

Eyes

The eye becomes cloudy almost to the point of whiteness and the fish lose vision. The causes behind this disease can vary. An increase in parasites in the aquarium is the most common cause but severe stress, old age, or malnutrition can all lead to this condition.

Treating this condition requires an investigation of water quality. Once the water quality is high enough, the fish will usually recover by themselves within 1–2 weeks. Thus, it is advisable to wait for 1–2 weeks before administering antibiotics.

Gills

Gill infection will cause breathing at the surface and fast respiration. Gill examination may reveal numbers of white spots. Wet mount of a gill biopsy may reveal mutifiliis trophozoites.

Treatment

Unfortunately, an efficient prevention of the disease by vaccination is not possible, although several studies identified potential vaccine candidate proteins, i.e. i-antigens, of the parasite.

Any treatment method must take into account the species of fish (some will not tolerate certain medications), how high the infection rate is, and the size and type of environment.

If it is detected before it becomes too serious, a number of different treatments can be applied. Only the free-swimming stage of the parasite is susceptible to treatment; neither the trophonts under the epithelium nor the tomont cysts can be killed.

Heat Treatment

Heat treatment can be highly effective, and it can be combined with other treatments. However, it can only be used on fish that can tolerate high water temperatures, and is unsuitable for cold water fish like koi and goldfish, but even in those cases, a higher water temperature will accelerate the life-cycle of the parasite, allowing other treatments to take effect sooner.

Chlorine

For treating koi & goldfish, chlorine, in the form of tapwater, is very effective in removing not only the threadlike parasites, but eventually the persistent cysts. Thread like infestations on fish will disappear overnight, cysts will take a couple of weeks and possibly a couple of water changes to eliminate. Aquarium lighting is used to detect the presence of parasites, as the filament like threads fluoresce at these light frequencies.

Salt

One method of treatment for ich consists of adding aquarium salt until a specific gravity of 1.002 g/cm^3 is achieved, as the parasites are less tolerant of salt than fish. This is not practical in ponds because even a light salt solution of 0.01% (100 mg/L; pure water at 4 °C/39 °F), would require large quantities of salt. Fish can be dipped in a 0.3% (3g/L; pure water at 4 °C) solution for thirty seconds to several minutes, or they can be treated in a prolonged bath at a lower concentration (0.05% = 500 mg (0,5g)/L; pure water at 4 °C).

Chemical Treatments

Chemical treatments include formalin, malachite green, chelated copper, copper sulphate, potassium permanganate and quinine. There are also a large number of proprietary treatments available for the treatment of white spot, and the related Oodinium (velvet disease). Chemical treatment is only effective against free-swimming juvenile parasites tomites. All treatments target the free-living theronts and tomonts, which only survive about two to three days in the absence of a host fish.

Prognosis

When Ich is diagnosed early, effective treatment is used, and stresses are minimised, mortality rates can be low. However, if the infection is at an advanced stage, treatment protocols are not followed, and the fish are stressed, higher death rates will occur. When a fish has had Ich eradicated, it may develop partial resistance to reinfection. Partially treated fish may initially harbour low numbers of unseen trophozoites, often in the gills. This subclinical carrier will cause another outbreak weeks later, most likely when stresses occur or uninfected fish are introduced to the aquarium.

Velvet (Fish Disease)

Velvet disease, also called gold dust disease is a fish disease caused by the dinoflagellate parasites of the genus *Piscinoodinium*, which gives the fish a dusty, slimy look. The disease occurs most commonly in tropical and (to a lesser extent) marine aquaria.

Life Cycle

The single-celled parasite's life cycle can be divided into three major phases. First, as a tomont, the parasite rests at the water's floor and divides into as many as 256 tomites. Second, these juvenile, motile tomites swim about in search of a fish host, meanwhile using photosynthesis to grow, and to fuel their search. Finally, the adolescent tomite finds and enters the slime coat of a host fish, dissolving and consuming the host's cells, and needing only three days to reach full maturity before detaching to become a tomont once more.

Pathology

Velvet (in an aquarium environment) is usually spread by contaminated tanks, fish, and tools (such as nets or testing supplies). There are also rare reports of frozen live foods (such as bloodworms) containing dormant forms of the species. Frequently, however, the parasite is endemic to a fish, and only causes a noticeable "outbreak" after the fish's immune system is compromised for some other reason. The disease is highly contagious and can prove fatal to fish.

Symptoms

Initially, infected fish are known to "flash," or sporadically dart from one end of an aquarium to another, scratching against objects in order to relieve their discomfort. They will also "clamp" their fins very close to their body, and exhibit lethargy. If untreated, a 'dusting' of particles (which are in fact the parasites) will be seen all over the infected fish, ranging in colour from brown to gold to green. In the most advanced stages, fish will have difficulty respirating, will often refuse food, and will eventually die of hypoxia due to necrosis of their gill tissue.

Treatment

Sodium chloride is believed to mitigate the reproduction of Velvet, however this treatment is not itself sufficient for the complete eradication of an outbreak. Additional, common medications added directly to the fish's environment include copper sulphate, methylene blue, formalin, malachite green and acriflavin, all of which can be found in common fish medications designed specifically to combat this disease. Additionally, because Velvet parasites derive a portion of their energy from photosynthesis, leaving a tank in total darkness for seven days provides a helpful supplement to chemical curatives. Finally, some enthusiasts recommend raising the water temperature of an infected fish's environment, in order to quicken the life cycle (and subsequent death) of Velvet parasites; however this tactic is not practical for all fish, and may induce immunocompromising stress.

Hexamita

Hexamita is a genus of parasitic diplomonads. It is related to *Giardia*. *H. columbae* and *H. meleagridis* lives in the intestines of birds. *H. muris* and *H. pitheci* lives in the intestines of mammals. *H. salmonis* and *H. truttae* lives in the intestines of fish.

The genus also includes the species *Hexamita inflata*.

It is believed that *Hexamita* parasites are one possible cause for head and lateral line erosion ("hole-in-the-head disease") in aquarium fishes.

Lymphocystis

Lymphocystis is a common viral disease of freshwater and saltwater fish. Aquarists often come across this virus when their fish are stressed such as when put into a new environment and the virus is able to grow.

The fish starts growing small white pin-prick like growths on their fins or skin and this is often mistaken for Ich/Ick (Ichthyophthirius multifiliis) in the early stages. It soon clumps together to form a cauliflower-like growth on the skin, fins, and occasional gills.

There is no known treatment for this virus, though some aquarists recommend surgery to remove the affected area if it is very serious.

Eventually the growths inhibit the fish's ability to swim, breathe or eat, and secondary bacterial infections usually kills the fish.

Usually the best cure is to simply give the fish a stress free life, a weekly bacteria treatment and the virus will slowly subside and the fins will repair themselves. This can take many months.

Flavobacterium Columnare

Flavobacterium columnare is a thin Gram-negative rod bacterium of the genus *Flavobacterium*. The name derives from the way in which the organism grows in rhizoidcolumnar formations.

The species was first described by Davis (1922), and the name was validated by Bernardet and Grimont (1989).

F. columnare can be identified in the laboratory by a five-step method that demonstrates:

1. the ability to grow on a medium containing neomycin and polymyxin B;
2. production of yellow pigmented rhizoid (root-like in appearance) colonies;

3. production of a gelatin-degrading enzyme;
4. binding of Congo red dye to the colony; and
5. production of a chondroitin sulphate-degrading enzyme.

The species has been known previously as *Flexibacter columnaris*, *Bacillus columnaris*, and *Cytophaga columnaris*.

F. columnare is one of the oldest known diseases among warm water fish, and manifests itself as an infection commonly known as Columnaris. Infections are the second leading cause of mortality in pond raised catfish in the southeastern United States.

The Quarantine

The goal of quarantine is to prevent problems in the main tank due to sickness. A quarantine tank should be used before to introduce any newly acquired animals in the main tank and to treat fish that are already sick. By doing this, the aquarist can avoid the spread of the disease and make it easier to treat the fish.

List of Freshwater Aquarium Invertebrate Species

This is a list of invertebrates, animals without a backbone, that are commonly kept in freshwater aquaria by hobby aquarists. Numerous shrimp species of various kinds, crayfish, a number of freshwater snail species, and at least one freshwater clam species are found in freshwater aquaria.

Crustaceans

Shrimp

- *Atyaephyra desmaresti*, Iberian/European dwarf shrimp
- *Atyoida pilipes* Green lace shrimp
- *Atyopsis moluccensis*, Bamboo Shrimp
- *Caridina cf. babaulti var. green*, green shrimp
- *Caridina babaulti var. malaya*, malayan dwarf shrimp
- *Caridina babaulti var. stripes*, striped shrimp
- *Caridina gracilirostris*, red nose shrimp
- *Caridina cf. gracilirostris*, white-nose Shrimp
- *Caridina multidentata*, Amano shrimp
- *Caridina cf. cantonensis var. bee*, bee shrimp
- *Caridina cf. cantonensis var. blue tiger*, blue tiger shrimp
- *Caridina cf. cantonensis var. crystal black*, crystal black shrimp

- *Caridina cf. cantonensis var. crystal red*, crystal red shrimp
- *Caridina cf. cantonensis var. tiger*, tiger shrimp
- *Caridina cf. propinqua*, tangerine shrimp
- *Caridina cf. serrata var. chinese zebra*, Chinese zebra shrimp
- *Caridina sp.*, Indian Dwarf Shrimp
- *Caridina sp.*, Indian Whitebanded Shrimp
- *Caridina woltereckae*, Sulawesi harlequin shrimp
- *Caridina dennerli*, Sulawesi cardinal shrimp
- *Caridina cf. breviata*, bumblebee shrimp
- *Caridina sp.*, black midget shrimp
- *Caridina serratirostris*, Ninja Shrimp
- *Caridina pareparensis parvidentata*, malawa shrimp
- *Caridina simoni simoni*, Sri Lankan Dwarf shrimp
- *Desmocaris elongata*, Guinea Swamp Shrimp
- *Euryrhynchus amazoniensis*, Amazon zebra shrimp
- *Halocaridina rubra*, Hawaiian Red shrimp (brackish, but can be acclimated to freshwater)
- *Macrobrachium assamensis*, Red claw shrimp
- *Macrobrachium dayanum*, Rusty longarm shrimp
- *Macrobrachium eriocheirum*, Fuzzy claw shrimp
- *Macrobrachium faustinum*, Caribbean longarm shrimp
- *Macrobrachium kulsiense*, Sand shrimp
- *Macrobrachium lanchesteri*, a.k.a. *Cryphiops lanchesteri*, Riceland prawn
- *Micratya poeyi*, Caribbean dwarf filter shrimp
- *Neocaridina heteropoda (wild type)*, Wild type cherry shrimp
- *Neocaridina heteropoda var. red*, Cherry shrimp
- *Neocaridina heteropoda var. yellow*, Yellow shrimp
- *Neocaridina palmata*, Marbled dwarf shrimp
- *Neocaridina cf. zhangjiajiensis var. blue pearl*, Blue pearl shrimp
- *Neocaridina cf. zhangjiajiensis var. white*, Snowball shrimp
- *Palaemonetes ivonicus*, amazon glass shrimp
- *Palaemonetes paludosus*, American ghost (glass, grass) shrimp
- *Palaemonetes varians*, European ghost shrimp

- *Potamalpheops sp.*, purple zebra shrimp
- *Triops australiensis*, Australian tadpole shrimp
- *Triops cancriformis*, horseshoe shrimp
- *Triops longicaudatus*, longtail tadpole shrimp
- *Xiphocaris elongata*, Yellow nose shrimp

Crayfish

- *Procambarus alleni*
- *Procambarus clarkii* Red Swamp Crayfish
- *Procambarus sp. Marmourkrebs* Marble Crayfish
- *Cherax destructor*
- *Cherax quadricarinatus*
- *Cambarellus montezuema*
- *Cambarellus shufeldtii*
- *Cambarellus patzcuarensis* Dwarf Crayfish
- *Pacifastacus leniusculus* Signal Crayfish

Branchiopods

- *Artemia franciscana*, San Francisco brine shrimp
- *Artemia salina*, Utah salt lake brine shrimp
- *Artemia nyos*, Sea Monkeys
- *Parartemia zietziana*, Sea Dragons
- *Artemia sp.*, Siberian brine shrimp
- *Lepidurus apus*, golden tadpole shrimp
- *Triops longicaudatus*, longtail tadpole shrimp
- *Triops australiensis*, Australian tadpole shrimp
- *Triops cancriformis*, European tadpole shrimp
- *Thamnocephalus sp.*, beavertail fairy shrimp
- *Streptocephalus sp.*, redtail fairy shrimp
- *Streptocephalus sp.*, dry lake fairy shrimp
- *Brachinecta sp.*, winter fairy shrimp
- *Daphnia*, water flea
- *Moina*, Japanese water flea
- *Diplostraca*, clam shrimp

Mollusca

Gastropods

- *Asolene spixi* (apple snail)
- *Marisa cornuarietis* (apple snail)
- *Planorbis* species (standard ramshorn snails)
- *Pomacea bridgesii* (apple snail)
- *Pomacea canaliculata* (apple snail)
- *Physa* species (bladder, tadpole and physa snails)
- *Lymnaeidae* (pond and melantho snails)
- *Lymnaea stagnalis* (great pond snail)
- *Planorbarius corneus* (Great Ramshorn)
- *Melanoides tuberculata* (Malaysian trumpet snail)
- *Melanoides granifera* (spike tailed trumpet snail)
- *Viviparus malleattus* (Japanese trapdoor snail)
- *Clithon corona* (horned nerite snail)
- *Neritina natalensis* (zebra nerite snail)
- *Vittina semiconica* (red onion or tire tracked nerite snail)
- *Neritina reclivata* (olive nerite snail)
- *Nepterion violaceus* "Red Lips" (Red lips neritine snail)
- *Septaria porcellana* (freshwater limpet)
- *Neritina* sp. (mosaic nerite snail)
- *Neritina* sp. (tribal nerite snail)
- *Brotia pagudola* (horned armour snail)
- *Pachymelania byronensis* (west African freshwater snail)
- *Neritina puligera* (dusky nerite)
- *Paludomus sulcatus* (bella snail)
- *Thiara cancellata* (hairy tower lid snail)
- *Taia naticoides* (piano snail)
- *Brotia herculea* (giant tower cap snail)
- *Planorbella duryi* (miniature red ramshorn snail)
- *Tylomelania* species (Tylo or Sulawesi snails)
- *Tylomelania sp. Poso* "Yellow"

- *Tylomelania sp. Poso* "Orange Flash"
- *Tylomelania patriarchalis* (black Sulawesi snails)
- *Cipangopaludina leucythoides* (tiger tuba snail)

Oceanarium

An oceanarium can be either a marine mammal park, such as Marineland of Florida, or a large-scale aquarium, such as the Lisbon Oceanarium, presenting an ocean habitat with marine animals, especially large ocean dwellers such as sharks.

Marine Mammal Parks

Marineland of Florida, one of the first theme parks in Florida, USA, started in 1938, claims to be "the world's first oceanarium", while Ocean Park in Hong Kong claims to be "the world's largest oceanarium".

Marineland of Florida was developed as *Marine Studios* near St. Augustine in Marineland, Florida, which was followed in Florida by Miami Seaquarium, opened in 1955 and in California by Marineland of the Pacific, opened in 1954 near Los Angeles, and Marine World, Africa USA, opened in 1968 near San Francisco.

SeaWorld San Diego was opened in 1964, developed by four fraternity brothers Milt Shedd, Ken Norris, David DeMott and George Millay. SeaWorld Aurora opened in 1970 near Cleveland, Ohio.

SeaWorld Orlando was opened in 1973. SeaWorld (San Diego, Aurora, Orlando) was sold to Harcourt Brace Jovanovich (a publishing company listed on the New York Stock Exchange) in 1976. They purchased Marineland of the Pacific in 1986 and closed the park.

They had opened SeaWorld San Antonio in 1988. In 1989 they sold SeaWorld (San Diego, Aurora, Orlando, San Antonio) to Anheuser-Busch, the world's largest brewer and owner of the Busch Gardens Safari Parks, for US$ 1.1 billion. In 2001, Anheuser-Busch sold the Ohio park which finally ceased its activities in 2004. In the capital of Kazakhstan is situated the only Oceanarium in Central Asia

World's Largest Marine Life Park

The Marine Life Park located within Resorts World Sentosa, (Sentosa, Singapore), is set to be the world's largest oceanarium, with 700,000 fish in 20 million gallons of water.

Resorts World Sentosa has made research and education cornerstones of this attraction, which aims to inspire, educate and enrich the understanding and protection of the oceans.

Marine Public Aquariums

Modern marine aquariums try to create natural environments. A host of marine animals swim together in the four-story cylindrical tank of the New England Aquarium in Boston, which opened in 1969. At the National Aquarium in Baltimore, which opened in 1981, a walkway spirals up through the centre of two gigantic cylindrical tanks, the Atlantic Coral Reef and the Open Ocean, which display sharks, sawfish, and other sea creatures. Since then, many new aquariums have sought even greater realism, often concentrating on local environments. The richly endowed Monterey Bay Aquarium in California, which opened in 1984, is an outstanding example.

Bibliography

Anderson, A. : *Fish Nutrition in Aquaculture*, Springer, Delhi, 2009.

Atre, P. K.: *Fish Genetics and Aquatic Environment*, Navyug, Delhi, 2008.

Balfour, F. M.: *A Monograph on the Development of Elasmobranch Fishes*: London, Macmillan, 1878.

Breder, C. M. and Rosen, D. E.: *Modes of Reproduction in Fishes: How Fishes Breed*: New York, Natural History Press, 1966.

Burgess, Peter: *Tropical Fishlopaedia: A Complete Guide to Tropical Fish Care*, Howell Books, NY, 2000.

Chhapgar, B. F.: *Fishes of India*, Oxford University Press, Delhi, 2008.

Collette B., and Facey D.: *The Diversity of Fishes*, Blackwell Publishing, UK, 2004.

Deka, Manab: *Fish Fermentation: Traditional to Modern Approaches*, New India Pub., Delhi, 2009.

Dubey, Pushpalata: *Fish Management and Aquatic Environment*, Daya, Delhi, 2006.

Edminster, C. : *Fish Ponds for the Farm*, Agrobios, Delhi, 2004.

Ferguson, H.W.: *Systemic Pathology of Fish*, Iowa State Press, Ames, Iowa, 1989.

Gill, Emlyn M.: *Practical Dry Fly Fishing*. New York: Charles Scribners & Sons, 1912.

Grizzle J.M.: *Anatomy and Histology of the Channel Catfish*, Auburn Printing Co, 1976.

Gunther E.: *Rainbow Fish*. Barron's Educational Series, NY, 2000.

Halver, J.E.: *Fish Nutrition*, Academic Press, New York, 1989.

Harvey, B.J. and B.J. Hoar: *The Theory and Practice of Induced Breeding in Fish*, International Development Research Centre, Ottawa, 1979.

Heda, Nilesh: *Fresh Water Fishes of Central India: A Field Guide*, Vigyan Prasar, 2009.

Hoar, W. S. and Randall, D. J.: *Fish Physiology*: London, Academic Press, 1969.

Hutchinson, Horace. G.: *Fly-Fishing in Salt and Freshwater*. London: John Van Voorst, 1851.

Kishore, Braj Prasad Singh: *Fish Genetic and Biotechnology*, Swastik, Delhi, 2008.

Klinke H. H.: *Fish Pathology*, T.F.H. Publications, Inc. Neptune City, NJ. 1973.

Kulkarni, G.K.: *Fisheries and Fish Toxicology*, APH, Delhi, 2006.

Kumar, Naresh Agarwal: *Fish Reproduction*, APH, Delhi,. 2008.

Lewbart G.A. : *Self-Assesment Color Review of Ornamental Fish*, Iowa State Press,1998.

Malvee, Sangeeta: *Fish Genetics*, SBS Pub, Delhi, 2008.

Marshall, N. B.: *The Life of Fishes*, Weidenfield and Nicholson, London, 1965.

Mori, Fumitashi: *Aquarium Fish of the World: The Comprehensive Guide to 650 Species*, Chronicle Books, UK, 1991.

Petr, Tomi: *Fisheries in Irrigation Systems of Arid Asia*, Daya, Delhi, 2007.

Pollard, D. A. *Freshwater Fishes of South-eastern Australia*. McDowell, R. M. Reed, Sydney. 1980.

Poppe T.T., *A Color Atlas of Salmonid Diseases*, Academic Press, London, 1996.

Puri, Neelima: *Fish Endocrinology*, Vista International Pub, Delhi, 2008.

Quick, James: *Fishing The Nymph*. New York: Ronald Press, 1960.

Rebelin W.E.: *The Pathology of Fishes*, The University of Wisconsin Press, UK, 1975.

Robertson, D.R.: *Fishes of the Tropical Eastern Pacific,* University of Hawaii Press, Honolulu. 1994.

Rosen, D. E.: *Modes of Reproduction in Fishes, How Fishes Breed*, Natural History Press, New York, 1966.

Sandford, Gina: *Aquarium Owner's Guide*, DK Publishing, London, 2000.

Sharma, O.P.: *Fisheries Extension and Administration*, Agrotech Pub, Delhi, 2011.

Shepherd, C.J. and H. Bromage : *Intensive Fish Farming*, Blackwell Scientific Publications, Oxford (UK), 1989.

Silva, De: *Fish Nutrition in Aquaculture*, Springer, Delhi, 2001.

Smith, R.I. : *Lights' Manual: Intertidal Invertebrates of the Central California Coast,* University of California Press, Berkeley. 1975.

Srivastava, C. B. L.: *Fish Biology*, Narendra, Delhi, 2008.

Srivastava, M.P.: *Natural History of Fishes and Systematics of Freshwater Fishes of India,* Narendra Publishing House, Delhi, 2002.

Stickney, R : *Culture of Nonsalmonid Freshwater Fishes,* CRC Press, Boca Raton, Florida (USA), 1986.

Stoskopf, M.K.: *Fish Medicine*, W.B. Saunders Co. 1993.

Timmermans, J.A. : *Textbook of Fish Culture: Breeding and Cultivation of Fish*, Fishing News Books, Farnham (UK), 1986.

Turton, John: *The Angler's Manual; or Fly-fisher's Oracle*. London: R. Groombridge, 1836.

Wright, Leonard M. Jr.: *Fishing the Dry Fly As A Living Insect-An Unorthodox Method*. New York: E. P. Dutton & Co., 1972.

Index

G

H

I

L

M

P

R

S

W

❑❑❑